STUDY GUIDE TO ACCOMPA

BIOLOGY

Science and Life

MICHAEL R. CUMMINGS
UNIVERSITY OF ILLINOIS AT CHICAGO

PREPARED BY

JAY M. TEMPLIN

WEST PUBLISHING COMPANY

MINNEAPOLIS/ST. PAUL

NEW YORK

LOS ANGELES

SAN FRANCISCO

Design Judith Mara Riotto

Cover Image James Carmichael / The Image Bank

WEST'S COMMITMENT TO THE ENVIRONMENT

In 1906, West Publishing Company began recycling materials left over from the production of books. This began a tradition of efficient and responsible use of resources. Today, 100% of our legal bound volumes are printed on acid-free, recycled paper consisting of 50% new fibers. West recycles nearly 27,700,000 pounds of scrap paper annually—the equivalent of 229,300 trees. Since the 1960s, West has devised ways to capture and recycle waste inks, solvents, oils, and vapors created in the printing process. We also recycle plastics of all kinds, wood, glass, corrugated cardboard, and batteries, and have eliminated the use of polystyrene book packaging. We at West are proud of the longevity and the scope of our commitment to the environment.

West pocket parts and advance sheets are printed on recyclable paper and can be collected and recycled with newspapers. Staples do not have to be removed. Bound volumes can be recycled after removing the cover.

Production, Prepress, Printing and Binding by West Publishing Company.

TEXT IS PRINTED ON 10% POST CONSUMER RECYCLED PAPER

Printed in the United States of America
03 02 01 00 99 98 97 96 8 7 6 5 4 3 2 1 0

ISBN 0–314–21333–3

CONTENTS

INTRODUCTION

This study guide is written to help you study the material presented in *Biology: Science and Life* and your introductory biology class. It is designed to function both as a guide to the text and to test you on its material after you have read and studied each chapter. To help you, this study guide offers the following sections.

Chapter Overview. This section provides a quick review of the chapter's major concepts.

Learning Objectives. These statements point out what you should be able to do if you have mastered the content of the chapter. To get the most out of it, you should read it over once before you start the chapter, and then keep going back to it while you read.

Chapter Outline. This section repeats the headings and subheadings of the chapter. In addition there are supporting statements that explain the information of the chapter in more depth.

Study Hints. For each chapter, suggestions are provided to further enhance your study of the information.

Key Terms. Vocabulary is important when you study biology. This list provides many of the key terms in the chapter and asks you to define them. Half the battle in conquering the subject of biology is mastering the terms. Not only should you know the definition of each term but you should also be able to use the terms. You should also know why each is significant.

Chapter Concepts. A concept is a unifying idea. You are asked to explain a few concepts in each chapter to understand some of the most important ideas. They require some thought and organization to answer. You can check your answers in the key at the end of the chapter.

Practice Test. Each study guide chapter ends with a series of review questions. They are provided in the four formats most commonly encountered on exams: true/false, multiple choice, matching, and fill-in-the-blank. Answering them provides a final check on your mastery of the content in the chapter. You can check your answers in the key at the end of the chapter.

Chapter 1

MODERN BIOLOGY AND ITS UNIFYING THEMES

CHAPTER OVERVIEW

This chapter describes life by explaining the fundamental characteristics that all living organisms share in common. For example, organisms have a cellular structure. Other important ideas are explained early in the chapter. Because millions of different species currently inhabit the Earth, biologists have produced a scheme to classify these organisms logically. Learning the properties of a kingdom, the broadest category of classification, is easier than learning the properties of each individual species in the kingdom. In addition, the physical and natural worlds consist of a hierarchy of levels of organization, ranging from atoms to the biosphere.

Science as an activity is often misunderstood, particularly by the nonmajor biology student in college. Although science is knowledge, it is also a process that generates that knowledge. The steps of scientific inquiry produce the hypotheses, theories, and principles, or organizing themes, that are relevant to help us understand the biological and physical world. Often these principles have predictive value. For example, understanding the principles of genetics can help someone predict the probability that a certain genetic disorder, such as Huntington disease, may be inherited.

Science can offer the information needed to make decisions in society. A biology student in college may someday make a decision about the environment. A background in biology can help that citizen to evaluate the consequences of building a nuclear power plant or removing a species in the community. Therefore science is related to the everyday issues of society.

LEARNING OBJECTIVES

After studying this chapter, you should be able to

1. Identify the basic characteristic that organisms share in common.
2. List the five kingdoms that place organisms in categories.
3. State the categories of classifying organisms in the correct hierarchical sequence.
4. State the levels of organization of life in the correct hierarchical sequence.
5. Explain the steps in the process of science that produce hypotheses, theories, and principles.
6. Explain how science is important for social, political, and ethical issues.
7. Describe the unifying themes of biology.

CHAPTER OUTLINE

Organisms Share Basic Characteristics

Organisms Are Made of Cells

1. The cell is the basic unit of life.
2. An organism is either one cell or a group of cells.

Energy Transfer Is Essential for Life

1. Energy for life originates from the Sun.
2. All living things acquire energy from plants, either directly or indirectly.

Organisms Maintain a Relatively Stable Internal Environment

1. By homeostasis living things maintain a relatively stable internal environment.

Response to Stimuli Is Essential for Survival

1. All living things respond to environmental stimuli.

All Organisms Are Capable of Growth and Reproduction

1. Living cells are formed from preexisting cells.
2. Reproduction can be sexual or asexual.

Adaptation and Evolution Have Changed Organisms

1. Some members of a population have variations that enhance their growth and survival.
2. Adaptive variations remain in a population and change it over time.

Organisms Can Be Grouped into Five Kingdoms

1. The five kingdoms are Monera, Protista, Plantae, Fungi, and Animalae.
2. The categories to classify organisms, from the broadest to the most specific, are: kingdom—phylum—class—order—family—genus—species.

Life Has Distinct Levels of Organization

1. The levels from broadest to most specific are: biosphere—ecosystem—community—population—organism—organ system—organ—tissue—cell—organelle—molecule—atom.

The Nature of Science and Scientific Inquiry

1. The hypothesis and theory are steps of scientific inquiry.

Science and Social Values

1. Science is knowledge.
2. Science is also a process that generates principles or organizing themes.
3. Science can be used in social, ethical, and political issues.

Unifying Themes

1. Science is a process that helps us understand our environment.
2. Genetics and evolution offer common threads to understand all of the subjects of biology.
3. Science is a human endeavor.

STUDY HINTS

1. Scan the entire textbook to gain an overview of all of the subjects of biology that you will be studying.
2. After reading over the textbook quickly, try to relate each subject of biology to genetics. For example, do genes provide control systems with the blueprint for their development? For evolution, can a population change unless there is raw material for that change? Are genetic variations the raw material?

CHAPTER REVIEW

Key Terms

Define these key terms from the chapter:

1. homeostasis

2. kingdom

3. class

4. genus

5. molecule

6. tissue

7. community

8. ecosystem

9. hypothesis

10. theory

Chapter Concepts

1. From personal experience, select one example of homeostasis currently working in your body. For example, at the moment your internal body temperature is approximately 37 degrees Celsius. How does your body respond if the external environment becomes too cold or too warm? How do these responses verify the concept of homeostasis?
2. Can you apply the steps of scientific inquiry to solve everyday problems in your life? For example, your hypothesis can be that laundry detergent A cleans your clothes better that laundry detergent B. How can you test this hypothesis? How extensive will this testing need to be to become a theory?

Practice Test

TRUE/FALSE

___ 1. All organisms are composed of more than one cell.

___ 2. Green plants capture energy from sunlight.

___ 3. Budding is a form of sexual reproduction.

___ 4. Asexual reproduction results from the fusion of sex cells.

___ 5. The species on the Earth are grouped into nine different kingdoms.

___ 6. Cellulose is found in the cell walls of plants.

___ 7. A family is a group of orders.

___ 8. For *Homo sapiens,* the term *sapiens* refers to the species name.

___ 9. A cell is a group of similar tissues.

___ 10. A community is a broader level of organization than a population.

___ 11. All ecosystems compose the biosphere.

___ 12. A hypothesis cannot be tested experimentally.

___ 13. A theory results from the extensive testing of a hypothesis.

___ 14. Huntington disease is a genetic disorder.

___ 15. Science alone can provide the answers to questions of social issues.

MULTIPLE CHOICE

16. Through molecules, cells have a ___________ basis.
    a. aesthetic
    b. chemical
    c. metaphysical
    d. philosophical
17. Energy from all living things originates from the
    a. air.
    b. soil.
    c. sun.
    d. water.
18. Because of homeostasis, if a person eats a candy bar the level of glucose in the blood will eventually
    a. reach a concentration of zero.
    b. remain too high.
    c. remain too low.
    d. remain relatively stable.
19. By plants responding to environmental stimuli,
    a. roots and shoots grow downward.
    b. roots and shoots grow upward.
    c. roots grow downward and shoots grow upward.
    d. roots grow upward and shoots grow downward.
20. Most organisms use ___________ reproduction.
    a. asexual
    b. homeostatic
    c. nonhomeostatic
    d. sexual

21. Through evolution, a population of organisms
    a. changes through time.
    b. does not produce mutations.
    c. does not reproduce.
    d. eliminates favorable variations.
22. Organisms on the Earth are grouped into ___________ different kingdoms.
    a. four
    b. five
    c. seven
    d. six
23. A cell wall is found in organisms of the kingdom
    a. Animalia.
    b. Fungi.
    c. Protista.
    d. Plantae.
24. Select the broadest category of classification.
    a. genus
    b. kingdom
    c. order
    d. phylum
25. Select the most specific category of classification.
    a. family
    b. genus
    c. order
    d. phylum
26. Select the smallest level of organization.
    a. atom
    b. cell
    c. molecule
    d. tissue
27. Select the broadest level of organization.
    a. cell
    b. organ
    c. organ system
    d. tissue
28. Only one kind of species is present in a(n)
    a. biosphere.
    b. community.
    c. ecosystem.
    d. population.

29. In science, another name for an idea is a(n)
    a. hypothesis.
    b. law.
    c. postulate.
    d. theory.
30. For *Drosophila melanogaster, Drosophila* is the ___________ name.
    a. family
    b. genus
    c. phylum
    d. species

COMPLETION

31. In all living organisms, ___________ are the basic units of life.
32. Either directly or indirectly, all living things depend on ___________ organisms for their diet.
33. Plants tend to bend ___________ a source of light.
34. If sex cells do not fuse in cell division, it is a form of ___________ reproduction.
35. In the scientific name for an organism, the ___________ name is not capitalized.
36. Molecules in a cell consist of ___________ .
37. All of the species living in an area comprise a ___________ .
38. To become a theory, a ___________ must be experimentally tested.
39. ___________ occurs among organisms by the selection of genetically controlled variations.
40. An inherited unit called a ___________ controls the formation of eyes in fruit flies.

ANSWERS

Key Terms

1. Homeostasis is the ability to maintain a relatively constant internal environment in the organism.

2. A kingdom is a group of closely related phyla and the broadest category used to classify organisms.
3. A class is a group of closely related orders and a subdivision of a phylum in classification.
4. A genus is a group of closely related species and a subdivision of the family in classification.
5. A molecule is a chemical unit and a group of atoms bonded together.
6. A tissue is a group of cells organized to perform a specific function.
7. A community is all of the species living together in a given area.
8. An ecosystem is a community plus the physical environment.
9. A hypothesis is an idea that can be experimentally tested.
10. A theory is a hypothesis that has been tested extensively by a variety of methods.

Chapter Concepts

1. If the external environment becomes cooler you shiver to generate heat. Much of your blood remains deep in your body to retain heat. If it becomes warmer you perspire and the blood comes to the surface of your body to promote the escape of heat. In either case the organism responds to maintain a relatively constant internal environment for temperature.
2. Test each detergent under identical conditions. You can only vary the kind of detergent. Other factors, such as water temperature and the amount of clothes washed, must be identical for the different detergents compared. The results must be evaluated objectively. Repeated testing must be conducted to develop a theory.

Practice Test

1. False	11. True	21. a	31. cells
2. True	12. False	22. b	32. plant
3. False	13. True	23. d	33. toward
4. False	14. True	24. b	34. asexual
5. False	15. False	25. b	35. species
6. True	16. b	26. a	36. atoms
7. False	17. c	27. c	37. population
8. True	18. d	28. d	38. hypothesis
9. False	19. c	29. a	39. evolution
10. True	20. d	30. b	40. gene

Chapter 2

THE ORIGIN AND CHEMISTRY OF LIFE

CHAPTER OVERVIEW

This chapter describes the chemical basis for life. Elements are the simplest substances of all matter. Elements occur in the environment as indivisible particles called atoms. In nature atoms seldom exist alone but tend to bond into molecules and compounds. These bonds include the covalent, ionic, and hydrogen bonds.

The molecules and compounds of life are constantly reacting to form new substances. The formation of new chemical bonds stores energy. The breakage of these bonds releases energy. These reactions of life often reach a chemical equilibrium.

Much of the chemistry of life occurs in water. Substances such as acids and bases exhibit their properties when dissolved in water. The pH scale is a measure of the acidity of a solution.

Several families of organic compounds have important biological properties. One family is the carbohydrates. Simple sugars such as glucose provide a quick source of energy to cells. Polysaccharides such as starch store energy. Cellulose, another polysaccharide, is a structural component of the cell wall. Among other families, lipids store energy efficiently and are important to structure. Proteins offer a variety of functions ranging from catalysis to regulation and transport. Nucleic acids encode genetic information. All of these families of molecules, with their important functions, had to be established structurally in cells for life to originate.

LEARNING OBJECTIVES

After studying this chapter you should be able to

1. Explain how atoms combine to form molecules and compounds.
2. Compare the several types of chemical bonds for similarities and differences.
3. Discuss the needed conditions and energy changes in chemical reactions.
4. List the important properties of water.
5. Compare the behavior of acids and bases when dissolved in water.
6. Compare the several kinds of carbohydrates for molecular size and functions.
7. Discuss the structure and functions of lipids and steroids.
8. Discuss the many different biological functions of proteins.
9. Describe the biological role of nucleic acids.
10. Outline some of the key steps leading to the origin of life.

CHAPTER OUTLINE

Matter, Elements, and Atoms

1. There is a chemical composition to all living matter.
2. Matter is composed of elements.
3. Elements occur in discrete particles called atoms

Chemical Elements Have Unique Properties

1. Atoms are the smallest components of matter.
2. The three kinds of subatomic particles are the protons, neutrons, and electrons.
3. The atomic number of an element is the number of protons.
4. The atomic weight of an element is the number of protons and neutrons in the nucleus.

Isotopes Are Different Physical Forms of Elements

1. Isotopes of an element differ in the number of neutrons in their atoms.
2. Oxidation is the loss of electrons from an atom.
3. Reduction is the gain of electrons by an atom.

Elements Form Compounds and Molecules

1. A compound is a combination of bonded atoms in a fixed ratio.
2. Some compounds are composed of molecules.

Ionic Bonds Are Formed by Charged Atoms

1. An ion forms when an atom gains or loses electrons.
2. An ionic bond is the electrostatic attraction between ions.

Covalent Bonds Involve Electron Sharing

1. A covalent bond is a shared pair of electrons between the outermost energy levels of atoms.
2. Electrons are shared unequally in a polar covalent bond.

Hydrogen Bonds Are Weak Interactions

1. A hydrogen bond is a weak interaction of two atoms, one of which is hydrogen.

Chemical Reactions Transfer Energy and Rearrange Atoms

Chemical Bonds Store Energy

1. The energy is measured in kilocalories.
2. Anabolic reactions form molecules that store energy.
3. Catabolic reactions break bonds and release energy.

Chemical Reactions Require Several Conditions

1. Reacting molecules undergo collisions.
2. The energy of activation must be present for a reaction to occur.
3. Chemical reactions can be exothermic or endothermic.

Reactions Can Reach an Equilibrium

1. An equilibrium is reached in a reaction when the opposing reaction rates are equal.

Energy-Producing and Energy-Requiring Reactions Are Linked

1. Biological reactions occur as steps of a metabolic pathway.
2. Catabolic reactions release energy stored in ATP.

Water, pH, and Life

The Properties of Water Are Critical to Living Systems

1. Water is the most common substance in cells.
2. As a solvent water dissolves polar substances.
3. Water is also cohesive and resists temperature changes.

pH, Acids, Bases, and Buffers

1. Acids are substances that add hydrogen ions to a solution.
2. Bases lower the hydrogen ion concentration in a solution.
3. Buffers control the hydrogen ion concentration and pH of a solution.
4. The pH is a scale from 0 to 14 that is a measure of acidity.

The Chemistry of Biological Systems

Organic Molecules Contain Carbon

1. Organic compounds are the carbon compounds of cells.

Carbohydrates Are Energy Sources and Structural Building Blocks

1. Monosaccharides are the simplest carbohydrates.
2. Disaccharides consist of two bonded monosaccharides.
3. Polysaccharides are long bonded chains of monosaccharides.

Lipids Are Structurally and Functionally Diverse

1. Triglycerides consist of three fatty acids bonded to glycerol.
2. Lipids serve as an energy reserve and build cell membranes.

Phospholipids Play Important Roles in Membrane Function

1. Phospholipids have important cell membrane functions.

Cholesterol and Sex Hormones Are Steroids

1. Cholesterol is a precursor to make steroid hormones.

Amino Acids Are Linked Together to Form Proteins

1. The sequence of amino acids is the primary structure of a protein.
2. Biological functions of proteins include catalysis, structural components, storage, protection, regulation, nerve impulse transmission, motion, and transport.

Enzymes Act as Catalysts in Biological Reactions

1. Enzymes are catalysts in biological reactions.
2. Enzymes combine with substrates and make them reactive.

Nucleic Acids Store and Transfer Genetic Information

1. Nucleotides are the subunits of nucleic acids.
2. DNA and RNA are nucleic acids.

Origin of Life: How Did It All Begin?

1. Life originated when autotrophy and metabolism were established.
2. As a first step complex organic molecules were produced from simple inorganic substances.
3. Genetic systems were eventually established in cells.

STUDY HINTS

1. Quickly read the questions and problems at the end of the chapter before studying the chapter. After studying the chapter answer the questions. Try this approach for several chapters.
2. Obtain a molecular model kit that allows you to build molecules from atoms. This can help you to visualize some of the bonding patterns discussed in the chapter.
3. If a lab section is part of your biology course, a firsthand study of the organic compounds will add to the information in the text. Which ones, for example, are insoluble in water?
4. Look ahead to Chapter 3 in the textbook. Is chemistry relevant to an understanding of cell structure and function?

CHAPTER REVIEW

Key Terms

Define these key terms from the chapter:

1. atomic number

2. isotope

3. oxidation

4. compound

5. covalent bond

6. kilocalorie

7. metabolism

8. exothermic

9. adhesion

10. buffer

11. macromolecule

12. monosaccharide

13. polysaccharide

14. glycoprotein

15. saturated fatty acid

16. peptide bond

17. substrate

18. nucleotide

19. DNA

20. proteinoid

Chapter Concepts

1. How is the chemistry of life essentially the chemistry of water? Review the properties of water and compare how the different kinds of compounds relate to water. Do they all dissolve in water? Are some insoluble in water?
2. Some families of organic compounds are important biologically. Are they important to human biology? One good example is to relate the carbohydrates, lipids, and proteins to nutrition.

Practice Test

TRUE/FALSE

___ 1. Matter is anything that has mass and occupies space.

___ 2. Silicon is the most abundant element found in the crust of the Earth.

___ 3. Carbon has an atomic number of six.

___ 4. Oxidation is the gain of electrons by an atom.

___ 5. If a sodium atom loses an electron, its charge becomes positive.

___ 6. Hydrogen bonds are strong and not easily broken.

___ 7. An oxygen atom can form two covalent bonds.

___ 8. Exothermic reactions give off energy to their surroundings.

___ 9. Catabolic reactions release energy.

___ 10. Bases are substances that add hydrogen ions to a solution.

___ 11. The human blood usually has an acidic pH.

___ 12. Glucose is a not disaccharide.

___ 13. Sucrose is a monosaccharide.

___ 14. Lactose is formed by the covalent bonding of glucose and galactose.

___ 15. Cellulose is a disaccharide.

___ 16. Saturated fatty acids do not have double bonds between their carbon atoms.

___ 17. Cholesterol is a steroid molecule.

___ 18. Nucleotides are the subunits of protein molecules.

___ 19. Enzymes are the substrates that catalysts act on in chemical reactions.

___ 20. Amino acids are the subunits of nucleic acid molecules.

MULTIPLE CHOICE

21. Select the most common element in the human body.
    a. carbon
    b. hydrogen
    c. nitrogen
    d. oxygen
22. An element has an atomic number of 8 and an atomic weight of 18. The number of neutrons in its nucleus is
    a. eight.
    b. eighteen.
    c. six.
    d. ten.
23. Which subatomic particle changes in the isotopes of an element?
    a. electron
    b. neutron
    c. positron
    d. proton

24. Each of the following choices can be lost in oxidation except a(n)
    a. electron.
    b. hydrogen atom.
    c. proton.
    d. water molecule.
25. Select the incorrect statement about compounds.
    a. Elements are combined in a fixed ratio.
    b. Many are composed of molecules.
    c. Sodium chloride is not an example of a compound.
    d. Water is an example of a compound.
26. Select the element that can form four covalent bonds.
    a. carbon
    b. hydrogen
    c. nitrogen
    d. oxygen
27. Select the correct statement about the water molecule.
    a. It contains three atoms.
    b. It is a nonpolar molecule.
    c. The hydrogen atoms are negative.
    d. The oxygen atom is positive.
28. Covalent bonds usually contain ____________ kilocalories.
    a. 10 to 30
    b. 40 to 50
    c. 80 to 100
    d. 160 to 200
29. Select the correct description of anabolic reactions.
    a. Chemical bonds are broken.
    b. Energy is released from bonds.
    c. Molecules are made.
    d. They are not part of metabolism.
30. Select the correct statement about chemical equations.
    a. Products are written on the left.
    b. Reactants are written on the right.
    c. They represent chemical reactions.
    d. They show only exothermic reactions.
31. Select the incorrect statement about water.
    a. Its adhesion produces capillary action.
    b. It does not resist temperature changes.
    c. It is a solvent.
    d. It is the most abundant compound in cells.

32. Select the substance with the most basic pH.
    a. ammonia
    b. black coffee
    c. rainwater
    d. seawater
33. A pH of 3 has a hydrogen ion concentration that is ___________ times as great as a pH of 6.
    a. 10
    b. 100
    c. 1000
    d. 10000
34. Select the carbohydrate with the largest number of atoms.
    a. glucose
    b. glycogen
    c. lactose
    d. sucrose
35. The chemical breakdown of maltose produces two molecules of
    a. cellulose.
    b. galactose.
    c. glucose.
    d. sucrose.
36. Select the element that is missing in a neutral fat or triglyceride.
    a. carbon
    b. hydrogen
    c. nitrogen
    d. oxygen
37. Select the element that is a steroid.
    a. cellulose
    b. insulin
    c. RNA
    d. testosterone
38. Which level of protein structure involves more than one polypeptide chain?
    a. primary
    b. quaternary
    c. secondary
    d. tertiary
39. Select the component that is not part of the nucleotide structure.
    a. amino acid
    b. base
    c. phosphate
    d. sugar

40. Select the molecule that was not in the reaction mixture of the apparatus studied by Miller in 1953.
    a. hydrogen
    b. methane
    c. starch
    d. water vapor

MATCHING

Match each term to its correct description. Each choice is used just once.

___	41. calcium	a. polysaccharide
___	42. estrogen	b. sugar in milk
___	43. keratin	c. table salt
___	44. kilocalorie	d. element
___	45. lactose	e. unit to express energy
___	46. sodium chloride	f. solvent molecule
___	47. starch	g. protein
___	48. water	h. steroid

COMPLETION

49. Elements are arranged in a ____________ table of elements according to their proton and electron content.

50. ____________ are atoms with the same number of protons but a different number of neutrons.

51. Oppositely charged atoms are attracted by ____________ attraction.

52. The energy of ____________ is the minimum amount of energy needed for a reaction to occur.

53. A chemical ____________ is represented by the double arrows of a chemical reaction.

54. ____________ is the force that holds the molecules of the same substance together.

55. ____________ absorb and release hydrogen ions in a solution to maintain its pH.

56. The three elements in glucose are carbon, hydrogen, and ___________ .

57. ___________ are molecules in which sugars are bonded to proteins.

58. ___________ is a regulatory protein than controls the concentration of glucose in the blood.

59. DNA has the form of a double ___________ .

60. ___________ cells originated from symbiotic associations with prokaryotes.

ANSWERS

Key Terms

1. The atomic number is the number of protons in an element.
2. A isotope is one of at least two atoms of an element that differ in their number of neutrons.
3. Oxidation is the loss of electrons from an atom.
4. A compound is a substance in which elements bond in a fixed ratio.
5. A covalent bond links atoms by electron sharing from electrons in their outermost energy shells.
6. A kilocalorie is 1000 calories and a unit to measure energy stored in chemical bonds.
7. Metabolism is the sum of all chemical reactions in an organism.
8. Exothermic relates to a chemical reaction that releases energy.
9. Adhesion is the ability of a substance to stick to another substance.
10. A buffer controls the hydrogen ion concentration and pH of a solution.
11. A macromolecule is a large molecule consisting of chemically bonded subunits.
12. A monosaccharide is a simple sugar and the subunit of large carbohydrate molecules.
13. A polysaccharide is a long, chain-like carbohydrate consisting of chemically bonded monosaccharides.
14. A glycoprotein is a sugar bonded to a protein.
15. A saturated fatty acid is a subunit of lipids and does not have any double bonds between the carbon atoms in its chain.
16. A peptide bond is a covalent bond connecting amino acids.
17. A substrate is the molecule that an enzyme chemically changes.
18. A nucleotide is a subunit of a nucleic acid.
19. DNA is deoxyribonucleic acid, a repository for genetic information.
20. A proteinoid is a polymer of amino acids that acts as a catalyst.

Chapter Concepts

1. Substances are either soluble or insoluble in water. Many important biological substances dissolve in water. Glucose, for example, could not dissolve in the blood if it were not water-soluble. Other substances, however, such as phospholipids have hydrophilic (water-attracting) and hydrophobic (water-repelling) ends. This allows these molecules to set up a membrane with respect to water. Biological membranes have important functions.
2. Carbohydrates are a source of energy in the organism. Glucose offers a quick source of energy. Glycogen stores extra calories that are not needed immediately. An athlete will consume a complex carbohydrate such as starch which is broken down into simple sugars. Lipids and proteins provide the raw materials to build many molecules of biological importance ranging from energy storage to structural components.

Practice Test

1. True	16. True	31. b	46. c
2. False	17. True	32. a	47. a
3. True	18. False	33. c	48. f
4. False	19. False	34. b	49. periodic
5. True	20. False	35. c	50. isotopes
6. False	21. d	36. c	51. electrostatic
7. True	22. d	37. d	52. activation
8. True	23. b	38. b	53. equilibrium
9. True	24. d	39. a	54. cohesion
10. False	25. c	40. c	55. buffers
11. False	26. a	41. d	56. oxygen
12. True	27. a	42. h	57. glycoproteins
13. False	28. c	43. g	58. insulin
14. True	29. c	44. e	59. helix
15. False	30. c	45. b	60. eukaryotic

Chapter 3

THE STRUCTURE OF CELLS

CHAPTER OVERVIEW

This chapter portrays cell structure and function. All living things consist of cells and their products. Cells are usually microscopic. Some organisms are unicellular. All others are multicellular. Prokaryotic cells are simple without membrane-bound organelles. Eukaryotic cells are complex with many membrane-bound organelles.

The plasma membrane regulates the passage of substances into and out of cells. Its unique chemical structure determines its capabilities. If particles can pass through this membrane, there are several means of cell transport: diffusion, osmosis, active transport, exocytosis, and endocytosis.

The eukaryotic nucleus is the cell control center. Its chromosomes contain the genetic blueprint for this control. The nucleolus inside the nucleus helps to make RNA. The other cytoplasmic organelles include the endoplasmic reticulum for synthesis and packaging, lysosomes for chemical digestion, mitochondria for cellular respiration, chloroplasts for photosynthesis, and cilia and flagella for movement. The cytoplasm contains a network called the cytoskeleton.

Some organisms have a one-celled structure. In multicellular organisms similar cells are organized into tissues. Tissues are integrated as organs. Related organs work together in the organ systems of the organism.

LEARNING OBJECTIVES

After studying this chapter, you should be able to

1. Outline the steps leading to the development of the cell theory.
2. Contrast the structure of prokaryotic and eukaryotic cells.
3. Explain the plasma membrane structure by the fluid mosaic model.
4. Compare the passive processes of cell transport. For osmosis relate the terms hypertonic, hypotonic, and isotonic.

5. Explain active transport.
6. Compare exocytosis and endocytosis.
7. Describe the structure and function of the nucleus.
8. Describe the role of the nucleolus.
9. Explain the structure and function of the membrane-bound compartments of the cytoplasm.
10. Explain the structure and function of the mitochondrion.
11. Explain the structure and function of the chloroplast.
12. Explain the structure and function of the cilia and flagella.
13. Describe the makeup of the cytoskeleton.
14. Compare the several kinds of cell-cell interactions.
15. Relate the role of the cell to the higher levels-of-organization.

CHAPTER OUTLINE

The Cell Theory Started a Revolution in Biology

1. All living things consist of at least one cell.

How the Cell Theory Developed

1. Schleiden and Schwann developed the cell theory.
2. Hooke studied cells through the microscope.

The Cell Theory and the Origin of Life

1. Cells cannot arise by abiogenesis.
2. Pasteur provided evidence for this idea.

Cell Size Is Partly a Problem of Physics and Mathematics

1. Most cells are microscopic.
2. Cell size is limited by the surface to volume ratio.

There Are Two Types of Cells: Prokaryotic and Eukaryotic

1. Organisms of Monera are prokaryotic.
2. Organisms of the other four kingdoms are eukaryotic.

Prokaryotic Cells Have a Simple Organization

1. Prokaryotic cells lack membrane-bound structures.

Eukaryotic Cells Are Larger and Have a Complex Internal Structure

1. Eukaryotic cells have a distinct nucleus.
2. Their cytoplasm is divided into membrane-bound compartments.

The Plasma Membrane: Life's Boundary

1. The plasma membrane is a barrier and regulates transport.

The Fluid Mosaic Model of Membranes

1. The membrane is mostly a phospholipid bilayer with hydrophilic and hydrophobic ends.

Membrane Components Have Many Functions

1. Membrane proteins function as transport channels.

Passive Transport and Osmosis Are Types of Diffusion

1. By diffusion molecules spread out to an area of lower concentration. It is passive.
2. Osmosis is the diffusion of water through a semipermeable membrane. Water enters a hypertonic environment.

Active Transport Can Move Molecules against a Concentration Gradient

1. By active transport particles are moved against the concentration gradient. It requires cell energy.

Exocytosis and Endocytosis Move Large Molecules across Plasma Membranes

1. By exocytosis cytoplasmic vesicles are transported outside the cell.
2. By endocytosis (i.e., phagocytosis) particles enter the cell.

Cell Coats and Cell Walls Offer Protection and Support

1. Plant cell walls surround their plasma membrane.
2. Some cells produce spores.

The Nucleus: A Genetic Information Center

1. The eukaryotic cytoplasm contains many organelles.
2. A nuclear envelope surrounds the nucleoplasm of the nucleus

The Nuclear Envelope Encloses and Separates the Nucleus from the Cytoplasm

1. The nuclear envelope is a double membrane with pores for transport.

Chromosomes Are Carriers of Genetic Information

1. Chromosomes are strands of DNA combined with proteins.
2. The chromosomes contain genetic instructions to control cell activities.
3. The chromosome number is constant in the body cells for a given eukaryotic species.

The Nucleolus Is the Site of Ribosome Synthesis

1. Nucleoli are inside the nucleus and are associated with ribosome production.

The Cytoplasm Is Divided into Compartments by Membranes

1. Organelles are classified into four categories.

The Endoplasmic Reticulum Synthesizes Proteins for Secretion

1. The channel-like ER can be smooth. It is rough if coated externally with ribosomes.
2. The rough ER makes proteins.
3. The smooth ER makes other substances and packages proteins.

Molecular Sorting Occurs in the Golgi Complex

1. The Golgi complex modifies and distributes materials made by the ER.
2. It also supplies membrane to other cell structures.

Lysosomes and Peroxisomes Are Cytoplasmic Disposal Sites

1. Lysosomes collect enzymes that carry out chemical digestion.
2. Peroxisomes contain oxidative enzymes.

Vesicles and Vacuoles Are Sites of Storage and Transport

1. Vesicles are membrane-bound spaces for transport
2. Vacuoles are membrane-bound spaces for storage.

Energy Conversion Occurs in Mitochondria

1. The cristae are the inner-membrane folds of the mitochondrion.
2. The mitochondrion is the site of cellular respiration.

Chloroplasts Convert Solar Energy into Food

1. Chloroplasts are found in plant cells.
2. The choloplasts have grana surrounded by the stroma.
3. The chloroplast is the site of photosynthesis.

Cilia and Flagella Provide Locomotion

1. Both of these organelles have a similar internal structure.
2. Each is anchored to the cell by a basal body.

The Cytoskeleton Gives Cells Shape and Support

1. Microtubules provide cell shape and chromosome movement.
2. Intermediate filaments anchor cells.
3. Microfilaments help to organize cell division.

Cell-Cell Interactions Coordinate Cell Function

1. Tight junctions join animal cells and act as a barrier.
2. Desmosomes cement together adjacent cells.
3. Gap junctions provide communication between the cytoplasm of adjacent cells.
4. Plasmodesmata penetrate the cell walls of adjacent plant cells.

Cells Are Organized into Tissues and Organs

1. A tissue is a group of similar cells that carry out a specific function.
2. An organ is a combination of tissues performing a function.
3. Related organs work in an organ system.

STUDY HINTS

1. Sketch a typical eukaryotic cell. Can you represent the organelles correctly by their structural details and locations in the cell?
2. On the lefthand side of a sheet of paper list the major parts and organelles of the eukaryotic cell. After studying Chapter 3, from memory write a brief description across from each structure.

CHAPTER REVIEW

Key Terms

Define these key terms from the chapter:

1. abiogenesis

2. prokaryotic

3. eukaryotic

4. diffusion

5. osmosis

6. hypertonic

7. hypotonic

8. isotonic

9. active transport

10. chromosome

11. ER

12. Golgi complex

13. lysosome

14. mitochondrion

15. chloroplast

16. cytoskeleton

17. tight junction

18. tissue

19. organ

20. organ system

Chapter Concepts

1. A major part of cellular function involves the transport of substances, particularly the transport of water. Can you think of some applications of this idea in everyday events of nature? For example, what problems of water balance does a fish face if it inhabits a saltwater environment? What causes

plants to dehydrate and wilt? What, in part, causes swelling in a region of the human body?

2. The cell offers a specific place for every structure. These structures are not independent but work together. Provide some examples how these structures work together. As one case, active transport requires energy. How is this energy furnished?

Practice Test

TRUE/FALSE

___ 1. The theory of abiogenesis stated that organisms originated from living material.

___ 2. Smaller cells tend to have higher metabolic rates than larger cells.

___ 3. Bacteria have a prokaryotic cell structure.

___ 4. The surface/volume ratio generally becomes smaller as cells become larger.

___ 5. A cell wall is present in plant cells.

___ 6. Phospholipid molecules have a hydrophilic end and a hydrophobic end.

___ 7. Diffusion is the movement of particles from an area of lower concentration to an area of higher concentration.

___ 8. By osmosis water diffuses across a membrane into a hypertonic solution.

___ 9. During endocytosis substances exit from a cell.

___ 10. Animal cells are surrounded by a cell wall.

___ 11. The nuclear envelope consists of a single membrane.

___ 12. Ribosomes are synthesized in the nucleolus.

___ 13. The smooth ER is the site of protein synthesis.

___ 14. Peroxisomes are membrane-bound vesicles.

___ 15. Vesicles are membrane-bound spaces.

___ 16. Cristae are folds of the inner membranes of chloroplasts.

___ 17. Chlorophyll is found in the stroma of chloroplasts.

___ 18. Through photosynthesis radiant energy is converted into the chemical energy of a carbohydrate.

___ 19. Tight junctions bind the nucleus to the cytoplasm of a cell.

___ 20. A tissue is a group of similar cells.

MULTIPLE CHOICE

21. Select the smallest unit of measurement of length.
    a. centimeter
    b. micrometer
    c. millimeter
    d. nanometer
22. Select the largest structure.
    a. bacterial cell
    b. frog egg
    c. mitochondrion
    d. virus
23. Select the incorrect description of a prokaryotic cell.
    a. It is enclosed by a plasma membrane.
    b. It contains a membrane-bound nucleus.
    c. Mitochondria not present.
    d. Lysosomes not present.
24. Plant and animal cells generally have each of the following structures in common except
    a. centrioles.
    b. cytoskeleton.
    c. endoplasmic reticulum.
    d. ribosomes.
25. Cell A has a 0.9% solute concentration. Cell B has a 0.9% solute concentration. Cell A is ____________ compared to Cell B.
    a. hyperosmotic
    b. hypertonic
    c. hypotonic
    d. isotonic
26. Which of the following statements does not describe active transport?
    a. It does not require energy.
    b. Molecules move against a concentration gradient.
    c. Molecules move from an area of lower to higher concentration.
    d. It is used by cells in the nervous system.
27. Which cell is most likely to carry out phagocytosis?
    a. liver cell
    b. skin cell
    c. red blood cell
    d. white blood cell

28. Cellulose composes the ___________ of the plant cell wall.
    a. inner lamella
    b. middle lamella
    c. primary cell wall
    d. secondary cell wall
29. Select the species with the highest chromosome number.
    a. corn
    b. dog
    c. fruit fly
    d. human
30. Select the organelle found in eukaryotic and prokaryotic cells.
    a. mitochondrion
    b. lysosome
    c. ribosome
    d. vaults
31. Select the species with the lowest chromosome number.
    a. frog
    b. gorilla
    c. onion
    d. tobacco
32. Lysosome function is disrupted in
    a. albinism.
    b. Huntington disease.
    c. polydactyly.
    d. Tay-Sachs disease.
33. Select a product of photosynthesis.
    a. carbon dioxide
    b. glucose
    c. radiant energy
    d. water
34. Grana are structures found in
    a. chloroplasts.
    b. mitochondria.
    c. ribosomes.
    d. vacuoles.
35. The basal body is a structure associated with
    a. cell walls.
    b. cilia.
    c. mitochondria.
    d. plastids.

MATCHING

Match each term to its correct description. Each choice is used just once.

___ 36. chloroplast	a. site of cellular respiration
___ 37. cilia	b. site of protein synthesis
___ 38. cytoskeleton	c. site of photosynthesis
___ 39. ER	d. contains digestive enzymes
___ 40. lysosome	e. contains the chromosomes
___ 41. mitochondrion	f. establish movement for the cell
___ 42. nucleus	g. can be rough or smooth
___ 43. ribosome	h. network of microtubules/microfilaments

COMPLETION

44. Robert Hooke used a ___________ to study cells.

45. The ___________ is a unit that is 1/1000 of a meter.

46. Bacteria belong to the kingdom ___________ .

47. The ___________ model describes the arrangement of lipids and proteins in cell membranes.

48. By ___________ large proteins are transported out of the cell.

49. ___________ consist of four to eight flattened sacs in plant cells.

50. ___________ junctions permit communication between the cytoplasms of adjacent animal cells.

ANSWERS

Key Terms

1. Abiogenesis states that life can form from nonliving material.
2. A prokaryotic cell is simple without membrane-bound organelles.
3. A eukaryotic cell is complex with many membrane-bound organelles.
4. By diffusion particles move passively from a region of higher concentration to a region of lower concentration.
5. Osmosis is the diffusion of water through a membrane permeable to water.
6. A hypertonic environment has a higher solute concentration (and lower water concentration) compared to another environment.

7. A hypotonic environment has a lower solute concentration(and higher water concentration) compared to another environment.
8. Isotonic refers to two environments that have identical solute (and water) concentrations.
9. Active transport is the movement of particles from a region of lower concentration to a region of higher concentration with the cell expending energy.
10. Chromosomes are structures of DNA and protein with the genetic blueprint.
11. The ER is a channel-like organelle for synthesis (protein synthesis in the rough ER) and packaging.
12. The Golgi complex is a series of stacked disks that modify and distribute materials made in the cell.
13. The lysosome is a baglike structure with enzymes for chemical digestion.
14. The mitochondrion is the site of cellular respiration where ATP is generated.
15. The chloroplast is the site of photosynthesis.
16. The cytoskeleton is a network throughout the cytoplasm for support and other functions.
17. Tight junctions connect animal cells and provide a barrier.
18. A tissue is a group of similar cells that perform a specific function.
19. An organ is a group of tissues working together.
20. An organ system is a group of related organs working together.

Chapter Concepts

1. A fish in saltwater is in a hypertonic environment. It faces the problem of losing water by osmosis. A plant wilts when its cells lose water. The human body is about two-thirds water. Swelling in the body, in part, is due to more blood moving to that area. Blood is about 50% water.
2. Look for interactions between cell parts. The mitochondria furnish the ATP for the energy needed for active transport. If a ribosome makes a protein, the ER packages it and transports it. The Golgi complex can form from the membrane of the ER and convert to a vesicle for transport of the protein out of the cell.

Practice Test

1. False	11. False	21. d	31. c	41. a
2. True	12. True	22. b	32. d	42. e
3. True	13. False	23. b	33. b	43. b
4. False	14. True	24. a	34. a	44. microscope
5. True	15. True	25. d	35. b	45. millimeter
6. True	16. False	26. a	36. c	46. Monera
7. False	17. False	27. d	37. f	47. fluid-mosaic
8. True	18. True	28. c	38. h	48. exocytosis
9. False	19. False	29. b	39. g	49. dictyosomes
10. False	20. True	30. c	40. d	50. gap

♑ Chapter 4 ♒

CELLS AND ENERGY: RESPIRATION AND PHOTOSYNTHESIS

CHAPTER OVERVIEW

This chapter sets forth the important principles of metabolism, the sum of all energy exchanges in the organism. These energy changes are enzyme-mediated and occur in a series of steps in metabolic pathways. The energy exchanges involve ATP.

In photosynthesis light energy is converted into the chemical energy of a sugar molecule. Molecular oxygen is liberated. The light reactions of photosynthesis liberate oxygen from water and produce hydrogen atoms. Hydrogen acceptors carry these atoms to the dark reactions to produce the sugar from the hydrogen and carbon dioxide. Bacteria of the ocean depths generate energy by chemosynthesis.

Overall, in cellular respiration the energy of a sugar, or other nutrient molecule, is used to generate ATP. In addition to sugar, oxygen is the other important reactant. Carbon dioxide and water are produced. For each glucose molecule, 36 ATP molecules are generated to perform useful work in the cell. The production of ATP from glucose occurs in a series of metabolic steps of glycolysis, the Krebs cycle, and the electron transport chain. This last event works in conjunction with chemiosmosis. Most of the generated ATP is from electron transport. Fermentation is an alternative to this complete metabolism if oxygen is not available to the cell. A small amount of ATP is made and, therefore, fermentation is not as efficient.

Metabolism contributes to the homeostasis of the organism. For example, it releases heat energy to control body temperature.

LEARNING OBJECTIVES

After studying this chapter, you should be able to

1. Explain the overview of metabolism in cells.
2. Describe the role of ATP in cell energetics.
3. State the events of photosynthesis in one sentence.
4. Outline the events of the light reactions of photosynthesis.
5. Outline the events of the dark reactions of photosynthesis.
6. Explain chemosynthesis briefly.
7. State the events of cellular respiration in one sentence.
8. Outline the steps of glycolysis.
9. Outline the steps of acetyl CoA formation and the Krebs cycle.
10. Explain the role of the electron transport chain and chemiosmosis in ATP formation.
11. Calculate the ATP yield per glucose molecule from the different parts of cellular respiration.
12. Explain briefly how other nutrients in addition to glucose enter metabolism for ATP yield.
13. Contrast the results of fermentation to those of cellular respiration.
14. Describe how metabolism is related to metabolism.
15. Explain how metabolism is regulated.

CHAPTER OUTLINE

Metabolism Is a Method of Energy Conversion

1. Energy is the ability to do work.
2. Energy can only be changed and is always conserved.
3. Energy changes in a system increase entropy.
4. Metabolism is the sum of all energy exchanges in an organism.

Biochemical Reactions and Metabolism

1. Metabolic changes occur as a part of a pathway.
2. Each metabolic step is mediated by an enzyme.

Metabolic Pathways Are Organized Sets of Reactions

1. Metabolic changes are organized efficiently in the cell.

ATP Transfers Energy Within Cells

1. Energy transfers involve ATP.
2. An ATP molecule has a three-part structure.
3. The formation of ATP from metabolism stores energy.
4. The conversion of ATP into ADP + P releases energy for metabolism.

Photosynthesis Converts Solar Energy into Food

1. The light reactions of photosynthesis covert light energy into chemical energy.
2. The dark reactions of photosynthesis covert the chemical energy into food molecules.

Light Is Converted into Chemical Energy

1. Photosynthesis occurs in the thylakoids of the chloroplast.
2. Thylakoids contain chloroplasts and chlorophyll a and b.
3. Chlorophyll traps light and passes energized electrons to electron acceptors in photosystems I and II. ATP is formed.

Energy Transfer Is Mediated by NADPH

1. NADP+ transfers electrons and protons from water in the light reactions as NADPH.

What the Light Reaction Accomplishes

1. Water is split to release protons and electrons to photosystem II.
2. The released protons are moved across thylakoid membranes to make ATP.
3. NADPH enters the dark reactions of photosynthesis.

The Dark Reaction Converts Chemical Energy Into Food

1. ATP and NADPH from the light reactions form RuBP.
2. 1 molecule of RuBp and 1 molecule of carbon dioxide make 2 molecules of PGAL.
3. PGAL is converted to a six-carbon sugar such as glucose.

Chemosynthesis: An Alternative to Photosynthesis

1. Bacteria in the ocean depths use hydrogen sulfide and oxygen to generate energy.

Converting Food into Energy Is a Three-Stage Process

1. Overall, glucose reacts with oxygen to form carbon dioxide and water plus ATP and heat.
2. The three steps are glycolysis, the Krebs cycle, and electron transport.

Glycolysis Converts Glucose to Pyruvate

1. By eight enzymatic steps one glucose molecule is converted into two molecules of pyruvate.
2. There is a net gain of 2 ATP molecules per glucose molecule.
3. Glycolysis does not require molecular oxygen.

The Krebs Cycle Completes the Breakdown of Glucose

1. Each pyruvate molecule is converted to one acetyl CoA molecule.

2. Acetyl CoA enters the Krebs cycle.
3. Two ATP molecules are produced per glucose molecule.
4. The chemical changes in the Krebs cycle release electrons that are carried to electron acceptors.

ATP Synthesis Is the Third Stage in Converting Food into Energy

1. Molecules such as NADH carry electrons to the electron transport chain along the inner surface of the mitochondrion.
2. Oxygen is the final electron acceptor in the chain.
3. As the electrons are passed along these receptors, protons are produced and more ATP is generated.

Chemiosmosis Is Used to Form ATP

1. Protons from electron transport are moved across the mitochondrial membrane to make the ATP.

Bacteria Use Phosphorylation to Generate ATP

1. Bacteria move protons made from electron transport across plasma membranes to make ATP.

Calculating the Energy Yield in Converting Glucose to ATP

1. A total of 36 ATP molecules are produced per glucose molecule entering cellular respiration.
2. Glycolysis generates a net gain of 2 ATP molecules.
3. The Krebs cycle makes 2 ATP molecules.
4. Electron transport generates 32 of the 36 ATP molecules.
5. By total ATP generation from glucose, cellular respiration is highly efficient.

Converting Fats and Proteins into Energy

1. Fats are metabolized to make CoA.
2. Most amino acids are metabolized to CoA.
3. The derivatives of fats and proteins are metabolized by the Krebs cycle and electrons transport to make ATP.

Fermentation: Energy Conversion in the Absence of Oxygen

1. Yeast cells carry out alcoholic fermentation.
2. If molecular oxygen is absent to the yeast cell, pyruvate is converted to acetaldehyde. Two ATPs are made.
3. Skeletal muscles carry out lactate fermentation.
4. If molecular oxygen is absent to the muscle cell, pyruvate is converted to lactic acid. Some ATP is made.
5. When oxygen becomes available to the cell, lactic acid can be converted back to pyruvate by the liver.

Homeostasis and Metabolism

1. The efficiency of cellular respiration is about 40%, as it stores much useful chemical energy.
2. Energy not captured in chemical bonds is released as heat used to control the body temperature of a plant or animal.

Metabolism Is Regulated in Several Ways

1. The conversion of starch to glucose in plants or glycogen to glucose in animals is reversible.
2. The direction of these reversible conversions depends on the needs of the cells of the organism.

STUDY HINTS

1. As with all of the chapters in the textbook remember to quickly study the questions and answers at the end of the chapter before studying the chapter. Then attempt to answer the questions in depth.
2. After studying the kinds of metabolism in detail from the chapter, try to explain each one in your own words. Try to give a basic overview as a beginning in each case and add to this outline. For example, start with a balanced equation of photosynthesis. Then explain where water intervenes (light reactions) and carbon dioxide comes into play (dark reactions) and so forth.

CHAPTER REVIEW

Key Terms

Define these key terms from the chapter:

1. entropy
2. metabolism
3. ATP
4. thylakoid
5. NADPH

6. light reactions

7. dark reactions

8. glycolysis

9. acetyl CoA

10. chemiosmosis

11. fermentation

12. homeostasis

Chapter Concepts

1. Photosynthesis is the single most important chemical process that occurs on the Earth. Can you support this statement? Consider its effect on the atmosphere and in feeding the human population.
2. Is lactic acid fermentation in skeletal muscles adaptive? What are its advantages? Can you cite any disadvantages?

Practice Test

TRUE/FALSE

___ 1. As the amount of disorder increases the amount of entropy decreases.

___ 2. Oxygen is a product of photosynthesis.

___ 3. Oxygen is a reactant of cellular respiration.

___ 4. Energy is absorbed in the ADP molecule when ATP is converted to ADP.

___ 5. Sugar molecules are formed in the light reactions of photosynthesis.

___ 6. Chlorophyll b has a high absorption of yellow and green light.

___ 7. NADPH functions as a storage form of energy.

___ 8. The dark reactions of photosynthesis occur on the inner surface of the thylakoid membrane.

___ 9. Light and dark reactions are present in chemosynthesis.

___ 10. Chemosynthesis occurs on the surface of the Earth's oceans.

___ 11. The steps of glycolysis depend on the presence of molecular oxygen.

___ 12. The Krebs cycle occurs in the inner compartments of the mitochondria.

___ 13. Carbon dioxide produced by the Krebs cycle enters the bloodstream in humans.

___ 14. The conversion of ADP into ATP stores energy in ATP.

___ 15. Carbon dioxide is the final electron acceptor for electrons released from the Krebs cycle.

___ 16. Chemiosmosis is essential for the formation of ATP.

___ 17. Mitochondria are the energy carriers in the bacterial cell.

___ 18. The harvesting of useful energy by cellular respiration is more efficient than the efficiency of gasoline engines in cars.

___ 19. Yeast cells carry out lactate fermentation.

___ 20. In animals, glycogen is converted into glucose immediately after a meal.

MULTIPLE CHOICE

21. Which type of energy is stored in a glucose molecule?
    a. chemical
    b. electrical
    c. mechanical
    d. radiant
22. The function of enzymes in metabolism is to
    a. block end-product formation.
    b. provide substrates.
    c. release cofactors.
    d. speed up the chemical reactions.
23. Carbon dioxide is a
    a. product in photosynthesis and respiration.
    b. product in photosynthesis and a reactant in respiration.

c. reactant in photosynthesis and respiration.
d. reactant in photosynthesis and product in respiration.

24. Select the group of organisms that cannot conduct photosynthesis.
    a. algae
    b. fungi
    c. green plants
    d. some bacteria
25. The specific site of photosynthesis resides in
    a. basal bodies.
    b. centrioles.
    c. cristae.
    d. thylakoids.
26. The greatest light absorption by chlorophyll b is with ___________ light.
    a. blue
    b. green
    c. orange
    d. red
27. One role of NADP is to
    a. block light.
    b. provide water.
    c. transport electrons.
    d. trap light.
28. The splitting of one water molecule releases ___________ hydrogen ions.
    a. five
    b. four
    c. three
    d. two
29. Which molecule stores energy in the most stable way?
    a. ATP
    b. FAD
    c. NADPH
    d. glucose
30. One molecule of PGAL has ___________ carbon atoms.
    a. five
    b. four
    c. three
    d. two
31. In glycolysis, one molecule of glucose forms ___________ molecules of pyruvate.
    a. eight
    b. five

c. six
d. two

32. In glycolysis there is a small net gain of molecules of
    a. ADP.
    b. ATP.
    c. glucose.
    d. oxygen.
33. Select the incorrect statement about the Krebs cycle.
    a. Acetyl CoA enters this process.
    b. It produces carbon dioxide.
    c. It receives a three-carbon molecule.
    d. It releases electrons to a series of acceptors.
34. The final electron acceptor that works with the Krebs cycle is
    a. carbon dioxide.
    b. glucose.
    c. oxygen.
    d. phosphate.
35. The respiration of one glucose molecule produces about ___________ ATP molecules.
    a. ten
    b. thirty-six
    c. twenty-eight
    d. twenty
36. Most of the ATP molecules from cellular respiration are generated by
    a. electron transport.
    b. fermentation.
    c. glycolysis.
    d. substrate phosphorylation.
37. Which molecule is the major source of nitrogenous waste in humans?
    a. amino acids
    b. fatty acids
    c. glucose
    d. water
38. Select the incorrect statement about fermentation.
    a. It chemically changes glucose.
    b. It makes ATP.
    c. It is performed by yeast cells.
    d. It requires oxygen.
39. A likely site of lactate fermentation in humans is
    a. blood.
    b. liver.

c. skeletal muscles.
d. skin.

40. After a meal most extra glucose in the blood is converted and stored in the
a. blood.
b. brain.
c. liver.
d. spleen.

COMPLETION

41. ___________ is the sum of all energy changes in an organism.

42. ATP consists of a ___________ with three bonded phosphate groups.

43. Thylakoids occur in ___________ among the different organelles in cells.

44. Sunlight causes the flow of electrons from photosystem II to photosystem ___________ .

45. In the dark reactions of photosynthesis, more than half of the PGAL made is converted to ___________ .

46. In the first step of glycolysis, glucose is converted to ___________ .

47. One molecule of glucose produces six molecules of the gas ___________ .

48. During chemiosmosis, ___________ flow across the membranes of the mitochondrion.

49. ___________ is the molecule that directly enters the Krebs cycle.

50. ___________ is the regulation of the internal environment.

ANSWERS

Key Terms

1. Entropy is a measure of disorder in a system. As the disorder increases the entropy increases.
2. Metabolism is the sum of all energy exchanges in a system.
3. ATP, adenosine triphosphate, is made by photosynthesis and cellular respiration to perform useful work in the cell.
4. Thylakoids are specialized membranes where photosynthesis occurs.
5. NADPH is an electron carrier in both photosynthesis and cellular respiration.
6. The light reactions of photosynthesis drive the formation of oxygen gas from

water and liberate hydrogen atoms from water that are carried by acceptors.

7. In the dark reactions of photosynthesis the hydrogens from the light reactions are used to make a sugar.
8. Glycolysis is the first eight metabolic steps for obtaining energy from glucose. Pyruvate is made. Oxygen is not required.
9. Acetyl CoA is made from pyruvate and enters the Krebs cycle.
10. Chemiosmosis takes the hydrogens from electron transport and moves them across mitochondrial membranes to make ATP.
11. Fermentation coverts pyruvate to a side product, such as lactic acid, when oxygen is not available to the cell. A small amount of ATP is made.
12. Homeostasis is the maintenance of a regulated internal environment of the organism. Metabolism contributes to this regulation.

Chapter Concepts

1. Photosynthesis removes carbon dioxide from the atmosphere. As combustion releases this gas, the role of photosynthesis produces a balance for this gas in the air. The release of oxygen by photosynthesis replaces the consumption of this gas by combustion and cellular respiration, thus establishing another balance. Photosynthetic organisms are the great producers of food on the Earth. Almost everything eaten is either a plant or an organism that has eaten a plant either directly or indirectly.
2. Lactic acid fermentation is adaptive as is offers some ATP generation to the organism in the absence of oxygen. In the process, however, there is a price to pay. This is the accumulation of lactic acid, which you will learn is the source of fatigue in the muscle. This must eventually be removed from the muscle.

Practice Test

1. False	14. True	27. c	40. c
2. True	15. False	28. d	41. metabolism
3. True	16. True	29. d	42. nucleotide
4. False	17. False	30. c	43. chloroplasts
5. False	18. True	31. d	44. I
6. False	19. False	32. b	45. RuBP
7. True	20. False	33. c	46. glucose-6-phosphate
8. False	21. a	34. c	47. carbon dioxide
9. False	22. d	35. b	48. protons
10. False	23. d	36. a	49. acetyl Co-A
11. False	24. b	37. a	50. homeostasis
12. True	25. d	38. d	
13. True	26. a	39. c	

ଓ Chapter 5 ଓ

CELL DIVISION

CHAPTER OVERVIEW

This chapter explores the different kinds of cell division. Cell division is a characteristic of most species. Unicellular eukaryotes reproduce by mitosis. In multicellular eukaryotes mitosis multiples the somatic cell number and accounts for growth, development, and cell replacement. In mitosis a diploid cell produces two diploid cells. In meiosis a diploid cell forms four haploid cells, usually gametes or sex cells. Prokaryotes can divide by binary fission.

The entire cell division entails mitosis and cytoplasmic division. These events are part of a cell cycle which also includes interphase. The details of mitosis with cytoplasmic division are summarized over the stages of prophase, metaphase, anaphase, and telophase. The chromosome number and combination of the dividing cells is held constant.

Meiosis begins with a diploid cell. Two consecutive divisions form four haploid cells. Chromosomes are reshuffled into many possible ways, promoting genetic variation. The fusion of haploid cells from different sexes yields a diploid cell which can then divide and develop by mitosis. Therefore, mitosis and meiosis complete a life cycle of an organism and contribute to the constancy of the chromosome number of a species.

LEARNING OBJECTIVES

After studying this chapter, you should be able to

1. Compare the differences between mitosis and meiosis in overview.
2. List the basic events of binary fission.
3. Describe the stages of interphase.
4. Describe the stages of mitosis.

5. Explain the significance of mitosis with respect to growth, cancer, and other biological phenomena.
6. Describe the stages of meiosis I.
7. Describe the stages of meiosis II.
8. Compare the assortment of chromosomes in meiosis and mitosis for similarities and differences.
9. Explain how meiosis contributes to genetic variability.
10. Relate mitosis and meiosis to the life cycle of an organism.

CHAPTER OUTLINE

Cell Division Is a Property of Living Systems

1. Unicellular organisms reproduce by cell division.
2. Multicellular organisms produce new cells by cell division.
3. In multicellular organisms mitosis produces somatic cells and meiosis produces gametes.

Some Cells Divide by Binary Fission

1. Prokaryotic organisms carry out binary fission.
2. Each daughter cell receives a complete set of genetic information and cytoplasmic components.

The Cell Cycle Describes the Life History of a Cell

1. The division of a cell consists of mitosis and cytokinesis.
2. The cell cycle consists of cell division and the events between divisions, called interphase.

Interphase Has Three Stages

1. In the G1 stage the cell grows and replaces cytoplasmic structures.
2. In the S phase DNA duplicates.
3. In the G2 stage the cell produces the molecules needed for mitosis.

Chromosomes Have a Characteristic Structure

1. In diploid cells of eukaryotes the chromosomes occur in homologous pairs.
2. The diploid chromosome number for humans is 46.
3. Each arm or copy (chromatid) of a chromosome is connected by a centromere.

Mitosis Has Four Stages

1. Mitosis consists of four stages: prophase, metaphase, anaphase, and telophase.

Duplicated Chromosomes Become Visible in Prophase

1. During interphase the genetic material appears as chromatin.
2. During interphase the long fibers of chromatin coil up and condense.
3. Each visible chromosome consists of two sister chromatids (copies) connected by a centromere.
4. The centrioles migrate to opposite poles of the cell and organize a mitotic spindle.
5. Near the end of prophase the chromosomes move toward a mitotic spindle. The nucleolus and nuclear membrane have broken down.

Chromosomes Become Aligned in Metaphase

1. Chromosomes attach by their centromeres along the spindle fibers at the equator.

Duplicated Chromosomes Separate in Anaphase

1. The centromeres connecting the sister chromatids divide.
2. The chromosomes derived from the sister chromatids are pulled toward opposite poles of the cell.
3. Therefore, a complete set of chromosomes is pulled toward each pole of the cell.

Cytoplasmic Cleavage Occurs in Telophase

1. A cleavage furrow forms along the equator of the cell.
2. The cell pinches into two daughter cells, each with about one-half of the original cytoplasm and with a complete set of chromosomes and genetic instructions.
3. Most of events of telophase are the reverse of telophase.
4. Each daughter cell of telophase enters prophase.

Mitosis in Organisms with Cell Walls

1. The plant cell divides into two daughter cells along the cell plate.
2. Centrioles do not exist in plant cells to coordinate this.

Mitosis Is Essential for Growth and Cell Replacement

1. The zygote divides by mitosis and continues to multiply the cell number in the organism.
2. Mitosis continues to replace cells and heals wounds.

Cancer and Control of the Cell Cycle

1. Cancer cells divide too rapidly and repeatedly.
2. The control mechanisms from G1 to S are disabled in cancer cells.

Cell Division by Meiosis: The Basis of Sex

1. Meiosis converts diploid cells (2n)into haploid (n) sex cells. Four haploid cells are produced.
2. In the 2n to n conversion, the members of each homologous chromosome pair are separated.
3. At the end of meiosis, each haploid cell receives only one copy of the chromosome from each homologous pair.
4. The chromosomes are reshuffled for genetic variation.

An Overview of Meiosis: Converting Diploid Cells into Haploid Cells

1. Meiosis consists of two consecutive divisions. Each round has four stages: prophase, metaphase, anaphase, and telophase.
2. The first meiotic division is reductional division. From a beginning diploid cell the chromosomes of each homologous pair are separated without centromeric division into separate daughter cells.
3. The first meiotic division produces haploid cells, as each of the two daughter cells receives one chromosome from each homologous pair of the beginning cell.
4. The second meiotic division produces haploid cells with centromeric division.

There Are Two Divisions in Meiosis

1. Meiosis I begins with diploid cells. Each diploid cell has undergone DNA replication.

Meiosis I Reduces the Chromosome Number

1. Each chromosome becomes visible and consists of two chromatids.
2. In prophase each homologous chromosome pair undergoes synapsis and forms a tetrad. A tetrad consists of four chromatids of the two homologous chromosomes.
3. The synapsed pairs of chromosomes move to the equator of the cell during metaphase. The members of a pair attach to the same spindle fiber on opposite sides of the equator.
4. During anaphase one chromosome (with two chromatids) separated from its homologous member (with two chromatids) without centromeric division.
5. Therefore, homologous chromosome pairs are separated into two different daughter cells.
6. Each daughter cell produced is haploid. Each chromosome still consists of two copies or chromatids.

Meiosis II Begins with Haploid Cells

1. In each cell, the chromatids of each chromosome are separated by centromeric division. They now exist as single copies of the chromosomes.
2. Four haploid cells are produced, each with a single copy of each chromosome.

Meiosis and Genetics

1. Meiosis maintains a constant chromosome number and contributes to genetic variation.

Meiosis Produces New Combinations of Genes in Two Ways

1. In meiosis I homologous chromosome pairs undergo independent assortment. This assortment reshuffles chromosomes into new combinations and increases genetic variation.
2. The variation by assortment increases as the number of homologous chromosome pairs increases.
3. Recombination of homologous chromosomes also contributes to genetic variation.

Sexual and Asexual Reproduction Are Different Evolutionary Strategies

1. Genetic variation promotes diversity in a population which assures survival for some members. Therefore, it is adaptive.
2. Asexual reproductive does not promote diversity.

Meiosis and the Life Cycle

1. The fusion of two haploid sex cells produces a diploid cell that restores the full chromosome number. A diploid organism grows by mitosis and can eventually produce haploid sex cells.
2. One example of this is the alternation of the diploid sporophyte with a haploid gametophyte in plants.
3. Meiosis and the rest of the life cycle maintains the chromosome number of the species.

STUDY HINTS

1. Sketch a cell at the top of a blank page with three pairs of homologous chromosomes. Each chromosome of this diploid cell must consist of two chromatids. Complete the sketch by showing how the chromosomes are assigned into the produced cells by mitosis (on the left) and meiosis I and II (on the right).

2. Divide a page into two halves. On the left side describe the major events for each of the four stages of mitosis. On the right side describe the major events for each corresponding stage in meiosis I. What are the similarities? What are the differences?

CHAPTER REVIEW

Key Terms

Define these key terms from the chapter:

1. somatic cell

2. gamete

3. mitosis

4. meiosis

5. cytokinesis

6. homologues

7. diploid

8. haploid

9. chromatid

10. centromere

11. zygote

12. reductional division

13. synapsis

14. recombination

15. sporophyte

Chapter Concepts

1. An organism with a diploid cell structure has a selective advantage during evolution over an organism with a haploid cell structure? Can you explain the basis for this advantage?
2. The second meiotic division has many similarities to the events of mitosis. What are they?

Practice Test

TRUE/FALSE

___ 1. Mitosis produces gametes.

___ 2. Mitosis multiplies the number of cells in multicellular organisms.

___ 3. Bacteria can reproduce by binary fission.

___ 4. Distinct chromosomes are not visible in a cell during interphase.

___ 5. In a haploid cell, chromosomes exist as homologous pairs.

___ 6. The normal diploid chromosome number in humans is 46.

___ 7. G2 is the last stage of interphase before a cell undergoes mitosis.

___ 8. The condensation of chromosomes during prophase increases their length.

___ 9. The chromatids of a chromosome are connected by a centriole.

___ 10. During metaphase of mitosis, chromosomes are aligned along the equator of the cell.

___ 11. Cytokinesis is the division of the nuclear membrane in the cell.

___ 12. A zygote divides by mitosis.

___ 13. Cells are replaced in the bone marrow by mitosis.

___ 14. Meiosis produces tetraploid cells.

___ 15. Centromeres normally do not divide in the first division of meiosis.

___ 16. DNA replication does not occur in the S phase of meiosis.

___ 17. Meiosis I produces triploid cells.

___ 18. Meiosis II begins with haploid cells.

___ 19. Meiosis produces new combinations of genetic information.

___ 20. In the alternation of generations of plants, the gametophyte is the diploid organism.

MULTIPLE CHOICE

21. A somatic cell of an organism has 80 chromosomes. Meiosis in this organism will produce cells with __________ chromosomes.
    a. 20
    b. 40
    c. 80
    d. 160
22. Cytokinesis involves the
    a. division of the cytoplasm.
    b. division of the nucleus.
    c. replication of chromosomes.
    d. replication of organelles.
23. During interphase, the centrioles are replicated during the __________ phase.
    a. G1
    b. G2
    c. R
    d. S
24. During interphase, DNA synthesis occurs during the __________ phase.
    a. G1
    b. G2
    c. R
    d. S
25. During interphase, the growth of the cell with preparation for cell division occurs during the __________ phase.
    a. G1
    b G2

c. R
d. S

26. A karyotype displays
   a. cells of the body.
   b. chromosomes.
   c. gametes.
   d. organelles.
27. During mitosis, the nuclear envelope disappears during
   a. anaphase.
   b. metaphase.
   c. prophase.
   d. telophase.
28. During mitosis, cytokinesis is completed during
   a. anaphase.
   b. metaphase.
   c. prophase.
   d. telophase.
29. During mitosis, the chromatids of the chromosomes begin to separate during
   a. anaphase.
   b. metaphase.
   c. prophase.
   d. telophase.
30. During mitosis, the nuclear envelope reforms during
   a. anaphase.
   b. metaphase.
   c. prophase.
   d. telophase.
31. During mitosis, the centrioles begin to migrate toward opposite poles of the spindle during
   a. anaphase.
   b. metaphase.
   c. prophase.
   d. telophase.
32. In a sex cell of an organism that has 18 chromosomes, the diploid chromosome number for this organism is
   a. 9.
   b. 18.
   c. 36.
   d. 72.
33. In a somatic cell of an organism, the chromosome number is 52. Meiosis produces cells with ____________ chromosomes.
   a. 16

b. 26
c. 52
d. 104

34. During interphase, the genetic material appears as
    a. centromeres.
    b. chromatin.
    c. distinct chromosomes.
    d. sister chromatids.
35. A cell plate forms during mitosis of a
    a. animal cell.
    b. bacterium.
    c. plant cell.
    d. virus.
36. A cell starts five consecutive mitotic divisions. This will produce ___________ cells.
    a. 8
    b. 16
    c. 18
    d. 32
37. The haploid cells of an organism normally has 16 chromosomes. A zygote of the this organism has ___________ chromosomes.
    a. 8
    b. 16
    c. 24
    d. 32
38. The synapsis of homologous chromosomes means that they
    a. disappear.
    b. duplicate.
    c. pair together.
    d. separate.
39. The formation of chiasmata means that homologous chromosomes
    a. crossover.
    b. duplicate.
    c. remain in the same cell.
    d. separate.
40. The diploid cells of an organism has seven pairs of homologous chromosomes. By meiosis it can produce ___________ different chromosomal combinations.
    a. 64
    b. 128
    c. 256
    d. 512

COMPLETION

Label each one of the statements as describing mitosis or meiosis I.

41. Crossing over occurs: ___________ .

42. Maintains the chromosome number in dividing body cells: ___________ .

43. Reduces the chromosome number by one-half in cells: ___________ .

44. Directs growth and development: ___________ .

45. Produces germ cells: ___________ .

46. Replaces dead and worn-out body cells: ___________ .

47. Forms tetrads: ___________ .

48. Referred to as reductional division: ___________ .

49. Disruption of this process produces cancer: ___________ .

50. Part of the cell cycle: ___________ .

ANSWERS

Key Terms

1. A somatic cell is a diploid body cell of a eukaryotic, multicellular organism.
2. A gamete is a haploid sex cell produced by meiosis.
3. Mitosis is cell division producing diploid cells.
4. Meiosis is cell division producing haploid cells.
5. Cytokinesis is cytoplasmic division of the cell when it divides.
6. Homologues are chromosomes of a pair in a diploid cell. They are the same size and shape.
7. A diploid cell has a pair of homologous chromosomes.
8. A haploid cell has only one chromosome of each kind. It may consist of two copies (chromatids) or one copy of that chromosome.
9. A chromatid is one of two copies connected by the centromere of a chromosome.
10. The centromere is where the two chromatids of a chromosome are connected.
11. The zygote is the fertilized egg.
12. Reductional division is the first meiotic division which separates pairs of homologous chromosomes without centromeric division. The cells produced have a haploid , halved or reduced, chromosome number.
13. Synapsis is the pairing of homologous chromosomes during meiosis I.
14. Recombination is the exchange of homologous chromosome parts during meiosis I.
15. A sporophyte is the diploid organism in the life cycle of a plant.

Chapter Concepts

1. Diploid cells that divide meiotically can produce many genetic variations through the assortment between homologous chromosome pairs. These genetic variations promote diversity in the population, a definite adaptation. Haploid cells lack this mechanism for assortment and adaptive variation.
2. Meiosis II is similar to mitosis except that a haploid set of chromosomes experiences centromeric division in meiosis II instead of a diploid set as in mitosis. The chromosome number is kept constant: diploid to diploid in mitosis and haploid to haploid in meiosis II.

Practice Test

1. False	14. False	27. c	40. b
2. True	15. True	28. d	41. meiosis I
3. True	16. False	29. a	42. mitosis
4. True	17. False	30. d	43. meiosis I
5. False	18. True	31. c	44. mitosis
6. True	19. True	32. c	45. meiosis I
7. True	20. False	33. b	46. mitosis
8. False	21. b	34. b	47. meiosis I
9. False	22. a	35. c	48. meiosis I
10. True	23. a	36. d	49. mitosis
11. False	24. d	37. d	50. mitosis
12. True	25. b	38. c	
13. True	26. b	39. a	

☙ Chapter 6 ❧

PRINCIPLES OF HEREDITY

CHAPTER OVERVIEW

This chapter presents the patterns of inheritance through a historical perspective. The results of Mendel's monohybrid and dihybrid crosses are summarized. Mendel's work generated the principles of segregation and independent assortment. These principles apply to understanding inheritance patterns in humans today. For example, in humans brachydactyly is caused by a dominant allele. Albinism is caused by a recessive allele.

Mendel's work discredited the theory of blending inheritance. Parents contribute equally by factors (genes) to the genotypes of the offspring. However, a dominant allele can mask a recessive allele in a heterozygote. Genotypes are expressed as phenotypes. The monohybrid cross (Ss X Ss) and dihybrid cross (SsYy X SsYy) each produce results in the phenotypes among the offspring. These results provide quantitative evidence for the basic principles and patterns of genetics.

LEARNING OBJECTIVES

After studying this chapter, you should be able to

1. Evaluate the theories of inheritance stated before Mendel's work.
2. Summarize the results of Mendel's monohybrid cross.
3. Summarize the results of Mendel's dihybrid cross.
4. Define the following terms: true-breeding, recessive, dominant, phenotype, genotype, gene, allele, homozygous, and heterozygous.
5. Explain the principle of segregation.
6. Explain the principle of independent assortment.
7. Solve genetic problems using Mendel's principles.
8. Explain how Mendel's principles are used in human genetics today.

CHAPTER OUTLINE

Heredity: How Does It Work?

1. Heredity compares the traits of parents to offspring for similarities and differences.
2. Historical theories of heredity, such as pangenesis, did not correctly explain the mechanism for heredity.
3. Mendel showed that units of heredity, now called genes, are transmitted from parents to offspring through the sex cells.
4. Mendel's work restructured several ideas about heredity.

Mendel's Experimental Approach Resolved Many Unanswered Questions

1. Mendel chose garden pea plants for his experiments. They offered many advantages for genetic crosses.
2. Mendel studied seven characters in the plants.
3. Mendel analyzed the quantitative results from large samples of offspring.

Crossing Pea Plants: The Principle of Segregation

1. A monohybrid cross involves studying one character.
2. Mendel started with true-breeding plants in a P1 generation. These parents had contrasting characters such as smooth x wrinkled seeds.

Results and Conclusions from Mendel's First Series of Crosses

1. From the P1 generation all of the F1 offspring showed only one of the parental traits, such as smooth seeds.
2. F1 members were crossed to produce F2 offspring.
3. In the F2 generation , about 75% of the offspring showed the same parental trait as the F1 plants, such as smooth seeds.
4. However, in the F2 generation about 25% of the offspring showed the other parental trait, wrinkled seeds.
5. Mendel concluded that the F1 trait (smooth seeds) was dominant to the other trait (wrinkled seeds).
6. Each parent in a genetic cross had a pair of factors (genes) for a trait. Each parent passed one of these two factors to an offspring.
7. The genotype is the genetic makeup, the factors, of an organism. The phenotype is the observed properties of the expressed genotype.

Inheritance of a Single Trait: The Principle of Segregation

1. Paired factors of a gene separate during the formation of sex cells. Therefore, only one factor of the pair is found in a gamete.
2. From an Ss genotype, a gamete receives the S factor or s factor. The alternate factors for a gene are called alleles.

3. In the F1 monohybrid cross, Ss X Ss, the different genetic recombinations in the F2 offpsring are: 1/4 SS to 2/4 Ss to 1/4 ss.
4. Ss organisms are heterozygous. SS or ss plants are homozygous and true-breeding.
5. SS and Ss plants develop the dominant phenotype, smooth seeds. ss plants develop the recessive phenotype, wrinkled seeds.
6. The 1/4 SS and 2/4 Ss offspring account 75% of the F2 offspring showing one parental characteristic.
7. The 1/4 ss (25%) offspring show the other characteristic.

More Crosses with Pea Plants: The Principle of Independent Assortment

Crosses with Two Traits

1. A dihybrid cross involves two sets of characters Crosses with Two Traits
2. Mendel compared smooth seeds to wrinkled seeds and yellow seeds to green seeds.

Methods, Results, and Conclusions

1. The P1 parents were true-breeding: SSYY (smooth, yellow) and ssyy (wrinkled, green).
2. All of the F1 offspring were SsYy and smooth, yellow.
3. F1 plants, heterozygous at each gene pair (SsYy) were crossed.
4. The results of the F2 offspring were about: 9/16 smooth, yellow; 3/16 smooth, green; 3/16 wrinkled, yellow; and 1/16 wrinkled, green.

The Principle of Independent Assortment

1. A genotype of SsYy produces four different kinds of gametes by independent assortment: SY, Sy, sY, and sy.
2. If two SsYy genotypes are crossed, a 4 X 4 Punnett square summarizes the genotypes and phenotypes of the F2 offspring.
3. In the F2 offspring: 9/16 have at least one copy of each dominant allele (S and Y), 3/16 have at least one copy of S and are yy, 3/16 have at least one copy of Y and are ss, and 1/16 are ss.
4. Mendel's work and principles of inheritance were rediscovered in 1900 and stimulated interest in genetics.

Mendelian Inheritance in Humans

1. Mendel's principles of segregation and independent assortment can be applied to other sexually reproducing organisms, including humans.
2. Among the offspring from an Aa X Aa genetic cross the genotypes are: 1/4 are AA, 1/2 are Aa, and 1/4 are aa.
3. From this monohybrid cross the phenotypes are 3/4 normal pigmentation and 1/4 albinism (aa, a recessive trait in humans).

4. A 9/16 to 3/16 to 3/16 to 1/16 phenotypic ratio is generated in a dihybrid cross in humans due to independent assortment of gene pairs (AaDd) and their fertilizations.

Pedigree Analysis in Human Genetics

1. It presents an orderly pattern of family information in a readable chart.
2. Parents and offspring are represented by symbols on the chart.
3. Knowledge of Mendelian inheritance is used in genetic counseling.

STUDY HINTS

1. Draw a Punnett square for the cross Aa X Aa. Without referring to the textbook, summarize the results of the genotypes and phenotypes among the offspring.
2. Conduct the same exercise as in #1, but for the genetic cross AaBb X AaBb. Again, summarize the results without referring to the textbook.

CHAPTER REVIEW

Key Terms

Define these key terms from the chapter:

1. gene
2. monohybrid cross
3. recessive
4. dominant
5. phenotype
6. genotype
7. allele

8. homozygous

9. heterozygous

10. dihybrid cross

Chapter Concepts

1. Can you relate the principle of segregation, involving genes, to the pattern of chromosome assignment in meiosis? Review the events of meiosis I.
2. Can you relate the principle of independent assortment to the pattern of chromosome assignment in meiosis? Once again, review the events of meiosis I.

Practice Test

TRUE/FALSE

___ 1. Genes are passed from parents to offspring through the blood.

___ 2. Mendel stated that each parent contributes one factor to the offspring for a given trait.

___ 3. Mendel studied fruit flies to learn about heredity.

___ 4. Peas have a very long growth period.

___ 5. In the peas studied by Mendel, smooth seed shape was dominant to wrinkled seed shape.

___ 6. A monohybrid cross involves studying only one character.

___ 7. Mendel's genetic crosses proved the idea of blending inheritance.

___ 8. The phenotype refers to the observed properties of a genotype.

___ 9. A dominant allele for a trait is symbolized by a capital letter.

___ 10. According to the principle of segregation, a gamete receives both factors for a trait.

___ 11. An allele is an alternative form of a gene.

___ 12. SS represents a heterozygous genotype.

___ 13. SsYy X SsYy represents a dihybrid cross.

___ 14. The least common phenotype among the offspring from a dihybrid cross of SsYy X SsYy is a recessive phenotype for both traits.

___ 15. The most common phenotype among the offspring from a dihybrid cross SsYy X SsYy is not a dominant phenotype for both traits.

___ 16. A genotype of SsYy produces two genetically different kinds of gametes by independent assortment.

___ 17. In humans, albinism is caused by a recessive allele.

___ 18. In a genetic cross of Aa X Aa, there is a 50% chance for a heterozygous offspring.

___ 19. A square represents females in a pedigree analysis.

___ 20. In humans brachydactyly is caused by a dominant allele.

MULTIPLE CHOICE

21. What Mendel called factors scientists now call
    a. cells.
    b. chromosomes.
    c. genes.
    d. sex cells.
22. Each of the following was used by Mendel in his genetic crosses except
    a. analysis of data.
    b. plants with many differing traits.
    c. self-fertilization by parents.
    d. small sample sizes among offspring.
23. Select the dominant trait studied in pea plants by Mendel.
    a. constricted pod
    b. green seed
    c. long stem
    d. terminal flower placement
24. Select the recessive trait studied in pea plants by Mendel.
    a. axial flower placement
    b. short stem length
    c. smooth seed shape
    d. yellow seed color
25. The genotypes of the P1 generation of Mendel's monohybrid cross were
    a. SS X SS.
    b. SS X Ss.

c. SS X ss.
d. ss X ss.

26. The genotypes of the F1 generation of Mendel's monohybrid cross were
a. SS X ss.
b. Ss X Ss.
c. Ss X ss.
d. ss X ss.

27. Among the phenotypes of the offspring in the F2 generation of Mendel's monohybrid cross, __________ of the offspring showed the recessive trait.
a. 3/4
b. 1/2
c. 3/8
d. 1/4

28. In Mendel's monohybrid cross, the genotype of all of the F1 offspring was
a. SY.
b. SS.
c. Ss.
d. Ss.

29. In Mendel's monohybrid cross, __________ of the F2 offspring had the genotype Ss.
a. 3/4
b. 3/8
c. 1/2
d. 1/4

30. In Mendel's monohybrid cross, __________ of the F2 offspring had the genotype SS.
a. 3/4
b. 1/2
c. 1/4
d. 3/16

31. In Mendel's monohybrid cross, __________ of the F2 offspring had the genotype ss.
a. 1/4
b. 1/6
c. 1/8
d. 1/10

32. In Mendel's dihybrid cross, the genotype of all of the F1 offspring was
a. SSYY.
b. SSYy.
c. SsYy.
d. Ssyy.

33. In Mendel's dihybrid cross, __________ of the F2 offspring developed the recessive phenotype for both traits.
    a. 9/16
    b. 1/2
    c. 3/8
    d. 1/16
34. In Mendel's dihybrid cross, __________ of the F2 offspring developed the dominant phenotype for both traits.
    a. 9/16
    b. 1/2
    c. 3/8
    d. 1/4
35. A pea plant with smooth, yellow seeds has the genotype
    a. SsYy.
    b. Ssyy.
    c. ssYy.
    d. ssyy.
36. A pea plant with wrinkled, yellow seeds has the genotype
    a. SsYy.
    b. Ssyy.
    c. ssYy.
    d. ssyy.
37. By segregation, a genotype of Rr produces __________ genetically different gametes.
    a. 1
    b. 2
    c. 3
    d. 4
38. Through independent assortment, a genotype of SsRr produces __________ genetically different gametes.
    a. 2
    b. 4
    c. 6
    d. 8
39. In the genetic cross Aa X Aa in humans, the chance to produce an offspring that will not develop albinism is
    a. 3/4.
    b. 1/2.
    c. 3/8.
    d. 1/4.

40. In the genetic cross Bb X Bb in humans, the chance to produce an offspring that will develop brachydactyly is
   a. 3/4.
   b. 1/2.
   c. 3/8.
   d. 1/4.

COMPLETION

41. __________ -breeding parents produce offspring that all have the same traits as the parent.

42. The __________ is the genetic makeup of the organism.

43. What biologists now call genes Mendel called __________ .

44. The genotype ss is homozygous __________ .

45. A pedigree __________ shows the genetic information of a family on a readable chart.

46. By genetic __________ predictions can be made for the chance of a trait in children from parents of known genotypes.

Answer with fractions for the following genetic crosses in humans:

47. In the genetic cross Dd X Dd, the chance to produce a deaf child is __________ .

48. In the genetic cross Cc x Cc, the chance to produce an offspring with cystic fibrosis is __________ .

49. In the genetic cross Tt X tt, the chance to produce an offpsring with thalassemia is __________ .

50. In the genetic cross BbHh X BbHh, the chance to produce an offspring that will not develop brachydactyly or Huntington disease is __________ .

ANSWERS

Key Terms

1. A gene is a unit of inheritance on the chromosome inherited with a certain probability. However, at the time of Mendel's work he did not know that chromosomes existed.

2. A monohybrid cross involves only one character.
3. A recessive trait is expressed only by the genotype ss.
4. A dominant trait is expressed by the genotype Ss or ss.
5. The phenotype is the observed character of the genotype.
6. The genotype is the genetic content, stated symbolically, (Ss), that is expressed as a phenotype.
7. An allele is an alternative form of a gene: S or s for seed texture.
8. A homozygous genotype has the two same alleles, SS or ss.
9. A heterozygous genotype is Ss. According to Mendel, the phenotype for this is dominant, as the dominant allele S masks the effect of the recessive allele, s.
10. A dihybrid cross involves studying two characters.

Chapter Concepts

1. Review meiosis I. Actually the chromosomes of a homologous pair segregate. Therefore, a gene pair on this homologous pair will also segregate into separate sex cells.
2. Review meiosis I. Independent assortment is between two pairs of homologous chromosomes. This increases genetic variation. One gene pair on one chromosome pair produces only two kinds of sex cells. Two gene pairs on two different chromosome pairs produces four kinds of combinations.

Practice Test

1. False
2. True
3. False
4. False
5. True
6. True
7. False
8. True
9. True
10. False
11. True
12. False
13. True
14. True
15. False
16. False
17. True
18. True
19. False
20. True
21. c
22. d
23. c
24. b
25. c
26. b
27. d
28. c
29. c
30. c
31. a
32. c
33. d
34. a
35. a
36. c
37. b
38. b
39. a
40. a
41. true
42. genotype
43. factors
44. recessive
45. analysis
46. counseling
47. 1/4
48. 1/4
49. 1/2
50. 1/16

Chapter 7

GENES AND CHROMOSOMES

CHAPTER OVERVIEW

This chapter explains patterns of inheritance that do not conform to Mendelian principles. For some traits dominance is incomplete. A gene may have more than two alleles in the population and codominance can occur. There are many polygenic traits. In addition, many genes have their expression modified by the external environment.

Genes are located on the chromosomes. Therefore, the patterns of gene inheritance follow the patterns of chromosome behavior and inheritance. This correlation explains Mendels's principles of segregation and independent assortment. Cytology experiments discovered this correlation which also explains sex determination, X-linked inheritance, gene linkage, and the construction of genetic maps.

Chromosomal abnormalities include aneuploidy and polyploidy. The distortions in chromosome numbers of these abnormalities can be traced to nondisjunction during meiosis.

LEARNING OBJECTIVES

After studying this chapter, you should be able to

1. Solve genetic crosses involving incomplete dominance.
2. Solve genetic crosses involving a gene with more than two alleles and codominance.
3. Contrast polygenic inheritance to inheritance from only one gene pair.
4. Explain how the external environment can affect heredity.
5. Explain the relationship between cytology and the principles of heredity.
6. Describe the chromosomal determination of sex.
7. Contrast sex-linked inheritance with inheritance patterns involving genes on autosomes.

8. Explain linkage.
9. Explain how linkage studies are used to construct genetic maps.
10. Compare the different kinds of chromosomal abnormalities.

CHAPTER OUTLINE

Variations on a Theme by Mendel

1. Some cases reveal that dominance is not always complete.
2. Some cases defy other Mendelian principles.

Incomplete Dominance Has a Distinctive Phenotype

1. The heterozygote (Rr) can have a phenotype (pink flowers) intermediate to its RR (red) and (white) rr parents.
2. An F1 cross of Rr X Rr yields a 1:2:1 (red:pink:white) phenotypic ratio among the F2 offspring.

Many Genes Have More than Two Alleles

1. There are three alleles in the population for human ABO blood types.
2. Blood type A has two genotypes (AA, AO).
3. Blood type B has two genotypes (BB, BO).
4. Blood type AB (genotype AB) and O (genotype OO) each have one genotype.

Codominant Alleles Are Both Expressed

1. A heterozygote fully expresses both alleles as in blood type AB.

Some Traits Are Controlled by More Than One Gene

1. Mendelian F2 offspring showed discontinuous variation in the 3/4 to 1/4 phenotypic ratio.
2. Many variations are continuous as with human height.
3. Continuous variations are the result of polygenic inheritance.
4. Continuous variations are quantified by measurement and increase in number for a trait as more gene pairs are involved.

Gene Expression Can Be Affected by External Factors

1. Campodactyly is one example in humans.
2. Pigment production is affected by temperature in Himalayan rabbits.
3. The expression of HD in humans is affected by age.

The Idea That Genes Are Located on Chromosomes

1. Cytology was conducted when Mendel performed his experiments.
2. Initially the chromosome patterns seen in cytology were not related to Mendel's principles.

Behavior of Genes and Chromosomes: Generation to Generation

1. At the turn of the twentieth century the chromosome theory of inheritance merged parts of cytology with the known principles of genetics.

Chromosomes Can Help Determine Sex

1. Autosomes are all of the chromosomes of a species other than the sex chromosomes.
2. In many species for the sex chromosomes the female is XX and the male is XY.
3. The XX or XY combination is determined in the zygote at fertilization.
4. Sex chromosome combinations in other species include XX (females) and XO (males) or ZW (females) and ZZ (males).
5. In humans the Turner syndrome is XO and the Klinefelter syndrome is XXY. These chromosomal differences arise from nondisjunction in meiosis.

Sex Linked Genes Have a Unique Pattern of Inheritance

1. A sex-linked gene is located on the X chromosome.
2. Studies with *Drosophila* and hemizygous males revealed new offspring ratios for the phenotypes of eye color.
3. X-linked conditions in humans include red-green color blindness and hemophilia.
4. Human X-linked traits are caused by recessive alleles and are much more common in males than females. A son often inherits the allele from his heterozygous mother.
5. By Y-linkage a trait is passed from father to son on the Y chromosome (i.e., TSD is humans).

Linkage and Gene Maps

1. The two gene pairs studied in Mendel's dihybrid cross were on different chromosome pairs. They will assort independently because the chromosomes do during meiosis.
2. Two gene pairs on the same chromosome pair will not assort independently.

Linked Genes

1. Linked genes are located on the same chromosome.
2. Studies with *Drosophila* showed that genetic crosses with two linked gene pairs do not yield Mendelian offspring ratios.
3. Crossing-over and recombination of the linked gene pairs in meiosis further modify the ratios.

Genetic Maps Can Be Constructed Using Crossover Information

1. The amount of crossing over between two linked gene pairs increases as the distance between them increases.

2. Information from genetic crosses with linked genes is used to construct genetic maps.

Chromosome Abnormalities

1. Nondisjunction and chromosomal breakage/realignment are the basis for chromosome abnormalities.

Changes in Chromosome Number

1. Aneuploidy is a small variation in the number of chromosomes of a species.
2. The Down syndrome is one example of aneuploidy in humans.
3. Polyploidy is a variation in whole sets of chromosomes of the species.
4. Polyploidy is common in plants.

Changes in Chromosome Structure

1. A deletion is a loss of a chromosome segment.
2. A translocation is a segment moved between chromosomes.
3. An inversion is a reversal in the gene order of a chromosome segment.

STUDY HINTS

1. Continue to preview the questions and problems at the end of the chapter. Answer all of these after studying the chapter in depth.
2. To better understand the principles of heredity, solve these genetic crosses. State the phenotypic ratios for the offspring.
   a. Rr X rr (flower color, incomplete dominance)
   b. AO X BO (human ABO blood type)
   c. Cc (on the XX human chromosomes) crossed with cY (human male)

CHAPTER REVIEW

Key Terms

Define these key terms from the chapter:

1. codominance

2. discontinuous variation

3. polygenic inheritance

4. cytology

5. sex chromosome

6. autosome

7. nondisjunction

8. linkage

9. hemizygous

10. genetic map

11. aneuploidy

12. polyploidy

13. deletion

14. duplication

15. translocation

Chapter Concepts

1. What does this chapter reveal about the continuing content of science? Is it constant or ever-changing?
2. Look ahead to the chapters 10 and 11 on evolution. How do you think the information in this chapter relates to this upcoming topic of evolution?

Practice Test

TRUE/FALSE

___ 1. Mendel's experimental results explained the pattern of incomplete dominance.

___ 2. Blood group AB in humans is an example of codominance.

___ 3. Skin color in humans is controlled by polygenic inheritance.

___ 4. Campodactyly is a congenital heart defect in humans.

___ 5. Huntington disease usually develops within two years after the birth of a child.

___ 6. Cytology is the study of cell structure.

___ 7. The X and Y chromosomes are autosomes.

___ 8. In humans an XX fertilized egg develops into a female.

___ 9. In birds a ZW fertilization develops into a female.

___ 10. The total number of chromosomes in the karyotype of the Turner syndrome is 45.

___ 11. The total number of chromosomes in the karyotype of the Klinefelter syndrome is 47.

___ 12. A sex-linked gene is located on an autosome.

___ 13. Red-green color blindness is an example of X-linked inheritance in humans.

___ 14. The Lesch-Nyham syndrome is caused by a recessive allele located on an autosome.

___ 15. In humans genes on the Y chromosome are inherited from the father.

___ 16. *Drosophila* is the genus name for the garden pea plant.

___ 17. Crossing-over occurs in the second meiotic division.

___ 18. Segregation of homologous chromosomes occurs in the first meiotic division.

___ 19. The total number of chromosomes in a karyotype of the Down syndrome is 45.

___ 20. Development of polyploid zygotes into adults is common in humans.

MULTIPLE CHOICE

21. Two plants with pink flowers mate. Assuming that pink develops from incomplete dominance, the phenotypic ratio for offspring of the next generation is
    a. 1:1.
    b. 3:1.
    c. 1:2:1.
    d. 1:1:1:1.
22. The chance to produce an offpsring with blood group O from two heterozygous parents with blood group A is
    a. 1/4.
    b. 1/2.
    c. 5/8.
    d. 3/4.
23. Which of the following choices does not describe human height?
    a. There is continuous variation in the population.
    b. It is the result of polygenic inheritance.
    c. The population distribution is a bell-shaped curve.
    d. It demonstrates Mendel's explanation for height.
24. Huntington disease is caused by
    a. a single dominant allele.
    b. a single recessive allele.
    c. many dominant alleles.
    d. many recessive alleles.
25. Cytology has proven that genes are located
    a. in the cytoplasm.
    b. on the cell membrane.
    c. on the chromosomes.
    d. outside the cell.
26. A normal human karyotype has ___________ autosomes.
    a. 2
    b. 23
    c. 44
    d. 46
27. In humans, an ___________ fertilization produces males.
    a. XX
    b. XXX
    c. XO
    d. XY

28. In birds, a __________ fertilization produces females.
    a. XXX
    b. XY
    c. YYY
    d. ZW
29. The number of autosomes in a karyotype of the Klinefelter syndrome is
    a. 44.
    b. 45.
    c. 46.
    d. 47.
30. A human gamete normally has __________ chromosomes.
    a. 22
    b. 23
    c. 24
    d. 25
31. *Drosophila* has __________ pairs of chromosomes in its body cells.
    a. 4
    b. 5
    c. 6
    d. 7
32. In fruit flies, if a white-eyed female mates with a red-eyed male, then among the offspring
    a. all of the females have red eyes.
    b. all of the females have white eyes.
    c. one-half of the females have red eyes.
    d. one-half of the males have white eyes.
33. Hemophilia in humans is caused by a
    a. dominant allele on an autosome.
    b. dominant allele on an X chromosome.
    c. recessive allele on an autosome.
    d. recessive allele on an X chromosome.
34. Protanopia is a form of
    a. color blindness.
    b. hair growth.
    c. muscle disease.
    d. nerve disorder.
35. Linked genes are
    a. in different cells.
    b. in different organisms.
    c. on the same chromosome.
    d. only on X chromosomes.

36. How does crossing over affect genetic variability in the gametes?
    a. It decreases it.
    b. It eliminates it.
    c. It has no effect on it.
    d. It increases it.
37. An example of a chromosome number in human body cells from aneuploidy is
    a. 45.
    b. 46.
    c. 92.
    d. 184.
38. An example of a polyploid chromosome number in humans is
    a. 46.
    b. 47.
    c. 48.
    d. 69.
39. A plant has a normal diploid chromosome number of 16. An example of a polyploid chromosome number is
    a. 8.
    b. 16.
    c. 18.
    d. 32.
40. A plant has a normal diploid chromosome number of 24. An example of an aneuploid chromosome number is
    a. 23.
    b. 24.
    c. 48.
    d. 96.

COMPLETION

41. In addition to heredity, ___________ affects the development of pigment production in Himalayan rabbits.
42. In addition to heredity, ___________ affects the determination of biological sex in alligators.
43. ___________ is the failure of homologous chromosomes to separate during meiosis.
44. If an organism carries only one copy of a gene instead of two it is not homozygous but it ___________ .

45. TDF is an example of ____________ -linked inheritance in humans.

46. As the ____________ between genes on a chromosome increases, the chance that they cross over increases.

MATCHING

Match each term to its correct description. Each choice is used just once.

___ 47. deletion	a. gene order on a chromosome is reversed
___ 48. duplication	b. loss of a chromosome segment
___ 49. inversion	c. extra copies of a chromosome are produced
___ 50. translocation	d. a segment is moved between two chromosomes

ANSWERS

Key Terms

1. In codominance a heterozygote fully expresses each allele.
2. In discontinuous variation there are distinct phenotypes from a genetic cross rather than an entire range.
3. By polygenic inheritance several gene pairs control a trait.
4. Cytology is the study of cell structure.
5. The sex chromosomes determine the sex of the organism
6. An autosome of a species is any chromosome other than the sex chromosomes.
7. During nondisjunction the chromosomes of a homologous pair do not separate during meiosis.
8. Linkage is the coupling of chromosome inheritance to a given gene.
9. Hemizygous refers to males that have only one copy of an X chromosome, and therefore, one gene for a trait.
10. A genetic map states the order and distance of genes on a chromosome.
11. Aneuploidy is a variation of only one or a small number of chromosomes from the normal number in a species.
12. Polyploidy is a variation of an entire set or sets of chromosomes in a species.
13. A deletion is a loss of a chromosome segment.
14. A translocation is a movement of a segment between chromosomes.
15. An inversion is a reversal in the gene order on a chromosome.

Chapter Concepts

1. Through scientific inquiry science is a body of knowledge that is constantly changing. Mendelian principles, for example, offered only a base for under-

standing inheritance. Further experimentation has built on that base to expand the field of genetics.

2. A population cannot change, or evolve, without a source of change. That source of change is genetic variation. Many alleles of a gene polygenic inheritance, independent assortment, and crossing over all increase that variation. You will learn that this variation is adaptive for a population.

Practice Test

1. False	14. False	27. d	40. a
2. True	15. True	28. d	41. temperature
3. True	16. False	29. a	42. temperature
4. False	17. False	30. b	43. nondisjunction
5. False	18. True	31. a	44. hemizygous
6. True	19. False	32. a	45. Y
7. False	20. False	33. d	46. distance
8. True	21. c	34. a	47. b
9. True	22. a	35. c	48. c
10. True	23. d	36. d	49. a
11. True	24. a	37. a	50. d
12. False	25. c	38. d	
13. True	26. c	39. d	

☙ Chapter 8 ❧

MOLECULAR BASIS OF INHERITANCE

CHAPTER OVERVIEW

This chapter reveals the chemical identity of a gene and how it works. Genes control the biochemistry of an organism. A gene is a DNA molecule which controls protein synthesis through an RNA molecule, mRNA. DNA can replicate itself. This allows a cell to reproduce its genetic structure. Therefore, when that cell divides each daughter cell receives a complete set of genetic instructions. When DNA makes mRNA by transcription it provides a message that controls the synthesis of proteins by translation at the ribosome. By controlling protein synthesis genes control cell structure through these molecules. In addition, genes control metabolism this way because all enzymes consist of proteins. Each step of a metabolic pathway is controlled by an enzyme.

When the flow of genetic information does not work properly in the cell, diseases develop at the molecular level. An altered DNA molecule changes the flow of events in transcription and translation. A changed mRNA makes a different protein. Often this change is as subtle as one base nucleotide (subunit) in the DNA and RNA and, therefore, one amino acid in the synthesized protein. A disease such as sickle cell anemia develops this way.

LEARNING OBJECTIVES

After studying this chapter, you should be able to

1. Explain the relationship between genes and proteins.
2. Describe the experiments that proved that DNA is the carrier of genetic information.
3. Outline the structure of the DNA double helix.

4. Explain how DNA makes DNA by replication.
5. Explain how DNA makes mRNA by transcription.
6. Explain how mRNA makes a polypeptide by translation.
7. State the key facts about the genetic code.
8. Explain how mutations develop at the molecular level.

CHAPTER OUTLINE

Genes and Proteins: Unresolved Questions

1. In the early twentieth century several unanswered questions remained about genetics.

Genes Control Metabolism

1. Alkaptonuria is inherited as a recessive Mendelian trait.
2. An acid waste product is excreted in this condition because the enzyme to metabolize it is not made.

The Relationship between Genes and Enzymes

1. Beadle and Tatum stated the one gene, one enzyme hypothesis. Also, a mutated gene will not make the enzyme and this blocks a step in a metabolic pathway.
2. A newer hypothesis is the one gene, one polypeptide hypothesis. One gene makes one polypeptide.
3. In alkaptonuria a mutated gene does not make the polypeptide of the enzyme necessary to metabolize a waste product.

DNA as a Carrier of Genetic Information

1. An unanswered question remains: What is a gene?

Transfer of Genetic Traits in Bacteria

1. Experiments by Griffith showed that dead S cells can be transformed into virulent R cells by receiving a transforming factor.
2. Later experiments showed that the transforming factor was DNA, the carrier of genetic information.

Reproduction in Bacterial Viruses Involves DNA

1. Experiments by Hershey and Chase showed that bacteriophages use DNA and not protein as the carrier of genetic information.

Watson, Crick, and the Structure of DNA

1. Watson and Crick built a model of the DNA molecule.
2. A linear DNA molecule is a chain of bonded nucleotides.
3. There are four different nucleotides, each with a different base: A, G, C, or T.

4. The sugar and phosphates of this polynucleotide are linked with the bases hanging free in an exact order.

DNA Is a Double Helix

1. The Watson-Crick model showed that the two polynucleotide strands are coiled around a central axis and united by a series of two possible base pairs: A to T and G to C.

DNA Is Coiled Together with Proteins to Form Chromosomes

1. Chromatin consists of DNA wound to histone proteins in nucleosomes.

DNA Replicating Relies on Base Pairing

1. DNA duplicates by semiconservative replication.
2. The DNA helix unwinds and each half synthesizes a new complementary other half.
3. DNA polymerase links complementary nucleotides together.
4. Each sister chromatid of a chromosome consists of a double-stranded DNA double helix.

How Is Genetic Information Stored in DNA?

1. Genetic information is stored in cells as the order of DNA base pairs.
2. A gene is hundreds or thousands of nucleotides in sequence.
3. Linear sequences of nucleotides are converted into linear sequences of amino acids in cells. The number of sequences is almost infinite.

The Flow of Genetic Information Is a Multistep Process

1. By transcription one unwound strand of DNA makes a single strand of RNA.
2. RNA is a single-stranded nucleic acid with bonded nucleotides consisting of the sugar ribose. There are four ordered bases in the nucleotides: A, G, C, and U.
3. The RNA transcript, mRNA, directs protein synthesis in the cytoplasm by translation.
4. A codon is a base triplet on mRNA.

Transcription Produces Genetic Messages

1. The promoter is the DNA base sequence at the beginning of a gene. It binds to RNA polymerase.
2. The sequence of DNA bases in one gene orders the bases in the transcribed mRNA. From DNA to RNA, the base pairing rules are A to U, C to G, G to C, and T to A.
3. A terminator is the DNA base sequence at the end of a gene.

Translation Requires the Interaction of Several Components

1. Each tRNA molecule delivers a particular amino acid at the ribosome.
2. The codons of mRNA attract tRNA anticodons and direct the correct placement and order of amino acids to make a polypeptide.
3. An anticodon recognizes a complementary codon by the following base pair rules between RNA molecules: A to U, G to C, C to G, and U to A.
4. The steps of translation are initiation (by a start codon), elongation, and termination (by a stop codon).
5. The ordered amino acids are linked by peptide bonds.
6. The polypeptide made by translation leaves the ribosome and is further modified for cellular function.

The Genetic Code: The Key to Life

1. The cell has 64 codons (base triplets) that relate to 20 amino acids.
2. Bases are read in triplets and a triplet specifies one amino acid.
3. 61 codons specify amino acids. There are three stop or terminator codons.
4. The genetic code is almost universal among all life forms.

Mutations Are Changes in DNA

1. At the molecular level a mutation is a change in the nucleotide sequence of DNA.

Mutant Proteins Can Produce Mutant Phenotypes

1. A change in one DNA gene base changes one mRNA base which can change one amino acid in the polypeptide translated.
2. An example of this change in the flow of genetic information is in the recessive gene for sickle cell anemia. One change in a DNA base and corresponding RNA base changes one amino acid in one polypeptide of hemoglobin, accounting for the development of sickle cell anemia.
3. Most mutations are harmful. However, they are also the raw material for natural selection and evolution.

STUDY HINTS

1. To understand how genetic information flows in a cell, try the following exercise. Start with a DNA base sequence of TTACCGCTA. State the base sequence of the new DNA made by this original DNA base sequence. Read all bases from left to right.
2. Using the newly-made DNA base sequence, write the sequence of mRNA bases that can be transcribed from this DNA. Continue to read bases from left to right. These nine mRNA bases represent three codons. Can you write the order of bases of the three tRNA anticodons that are complementary to them?

CHAPTER REVIEW

Key Terms

Define these key terms from the chapter:

1. enzyme

2. bacteriophage

3. polynucleotide

4. chromatin

5. semiconservative replication

6. transcription

7. translation

8. codon

9. anticodon

10. mutation

Chapter Concepts

1. What is a gene? This is an important idea in modern biology. Emphasize how it works in the cell.
2. Compare the modern idea of a gene, which emphasizes chemical events in a cell, with the Mendelian interpretation of this unit of heredity. Are there similarities? Are there differences?

Practice Test

TRUE/FALSE

___ 1. Enzymes are the substrates that enter into chemical reactions.

___ 2. For *Streptococcus pneumoniae, Streptococcus* is the genus name for this bacteria.

___ 3. Griffith showed that the T strain of *Streptococcus pneumoniae* caused a disease in mice.

___ 4. A bacteriophage is a type of bacterium.

___ 5. Phages are composed of only DNA and protein.

___ 6. Amino acids are the subunits of nucleic acids.

___ 7. In the DNA double helix G pairs with C.

___ 8. In the DNA double helix A pairs with T.

___ 9. Chromatin consists of DNA and protein.

___ 10. Semiconservative replication refers to how DNA makes RNA.

___ 11. In eukaryotes RNA makes DNA which makes protein.

___ 12. By translation in eukaryotes RNA directs the synthesis of DNA.

___ 13. By transcription DNA directs the assembly of RNA.

___ 14. A codon consists of three nucleotides.

___ 15. The end of a gene is marked by a terminator region.

___ 16. The anticodon is part of the messenger RNA molecule.

___ 17. The function of transfer RNA is to bond the nucleotides of DNA.

___ 18. Elongation is a step in translation.

___ 19. Sickle cell anemia is an autosomal dominant disorder.

___ 20. Normal beta hemoglobin differs from sickle cell hemoglobin through the identity of at least ten amino acids.

MULTIPLE CHOICE

21. Metabolic pathways are best described as
    a. birth defects.
    b. chains of reactions.
    c. transport patterns on the plasma membranes.
    d. tubules in the cells.

22. Alkaptonuria is diagnosed by examining a
    a. brain scan.
    b. skin surface.
    c. temperature reading.
    d. urine sample.
23. Beadle and Tatum proposed the one gene, one ___________ hypothesis.
    a. cell
    b. enzyme
    c. gene
    d. mutation
24. Griffith proved that R strains of *Streptococcus pneumoniae* can cause a disease in mice when mixed with
    a. blood from the mice.
    b. cells from the mice.
    c. heat-killed R strains.
    d. heat-killed S strains.
25. A bacteriophage reproduces in
    a. *E. coli.*
    b. human cells.
    c. protozoa.
    d. viruses.
26. Which subunit is not a part of the DNA nucleotide?
    a. amino acid
    b. base
    c. sugar
    d. phosphate
27. In the DNA double helix, a base sequence of ATAGTC bonds with a base sequence of
    a. ATAGTC.
    b. AATGCA.
    c. GTCATC.
    d. TATCAG.
28. The base not found in DNA is
    a. A.
    b. C.
    c. G.
    d. U.
29. DNA polymerase is involved in
    a. conjugation.
    b. replication.
    c. transformation.
    d. transcription.

30. ___________ different amino acids are found in protein structure.
    a. 20
    b. 80
    c. 120
    d. 250
31. A codon is a structure found in
    a. glucose.
    b. mRNA.
    c. tRNA.
    d. proteins.
32. A base triplet of UUU on mRNA will bond to an anticodon of ___________ on tRNA.
    a. AAA
    b. GGG
    c. TTT
    d. UUU
33. RNA polymerase is involved in
    a. conjugation.
    b. replication.
    c. transformation.
    d. transcription.
34. An anticodon of UCA bonds to a codon of
    a. AGT.
    b. AGU.
    c. UCA.
    d. TCA.
35. Which of the following statements does not describe tRNA?
    a. It functions at the ribosome.
    b. It is a molecule of about 80 nucleotides.
    c. It is single-stranded.
    d. It works as an enzyme.
36. Which of the following is not a translation step?
    a. conservation
    b. elongation
    c. initiation
    d. termination
37. A peptide bond unites
    a. amino acids.
    b. fatty acids.
    c. monosaccharides.
    d. nucleotides.

38. A sequence of nine RNA bases would probably order ___________ amino acids at the ribosome.
    a. 3
    b. 6
    c. 9
    d. 12
39. A and P are sites on the
    a. ER.
    b. nucleus.
    c. plasma membrane.
    d. ribosome.
40. Select the codon that is not a stop codon.
    a. UAA
    b. UAG
    c. UGA
    d. UUU

COMPLETION

41. Name three genetic disorders produced from metabolic blocks in the metabolic pathway that accepts and changes phenylalanine: ___________ , ___________ , and ___________ .
42. Griffith named the ___________ factor.
43. The nucleotide bases in DNA are purines and ___________ .
44. Nucleosomes consist of DNA molecules and proteins called ___________ .
45. The number of mRNA codons is ___________ .
46. Name the three bases of the start codon: ___________ , ___________ , and ___________ .
47. A ___________ is a change in the nucleotide sequence of DNA.
48. The molecule ___________ is affected in sickle cell anemia.

ANSWERS

Key Terms

1. An enzyme is a catalyst that acts in a step of a metabolic pathway.
2. A bacteriophage is a virus that infects a bacterium.
3. A polynucleotide is a bonded series of nucleotides.

4. Chromatin is DNA molecules bonded with histone proteins in the nucleus.
5. By semiconservative replication a DNA double helix unwinds and each half makes a complementary new half. When finished, two double helices are produced.
6. By transcription one strand of DNA directs the synthesis of a linear RNA molecule.
7. By translation mRNA directs protein synthesis at the ribosome.
8. A codon is a base triplet on mRNA.
9. An anticodon is the base triplet on tRNA that attracts to a complementary codon during translation.
10. At the molecular level a mutation is a change in the nucleotide sequence of a DNA molecule.

Chapter Concepts

1. A gene is a DNA molecule that makes a message (mRNA) that directs the synthesis of a polypeptide (proteins). This is a viewpoint that emphasizes the functional role of a gene in metabolism.
2. Mendel viewed the gene as a factor that is inherited with a certain probability. His work did not uncover how a genotype is translated into a phenotype or, in other words, how a gene acts in the cell. A modern view, however, combines these two viewpoints, The recessive gene (Mendelian term) for sickle gene anemia has been analyzed at the molecular level.

Practice Test

1. False	13. True	25. a	37. a
2. True	14. True	26. a	38. a
3. False	15. True	27. d	39. d
4. False	16. False	28. d	40. d
5. True	17. False	29. b	41. PKU
6. False	18. True	30. a	42. transforming
7. True	19. False	31. b	43. pyrimidines
8. True	20. False	32. a	44. histones
9. True	21. b	33. d	45. 64
10. False	22. d	34. b	46. AUG
11. False	23. b	35. d	47. mutation
12. False	24. d	36. a	48. hemoglobin

☙ Chapter 9 ❧

RECOMBINANT DNA AND GENETIC ENGINEERING

CHAPTER OVERVIEW

This chapter unveils the many exciting advances in genetics of the late twentieth century. Recombinant DNA technology can produce cells with newly inserted genes and new properties. These genetically engineered cells can be cloned and used to solve numerous problems in society. Applications include genetic mapping, diagnosing and treating human birth defects, constructing the human genome, and generating more productive crops and domesticated animals.

This advancing field of science is also accompanied with controversy. How will these advances be used? Many ethical, social, and legal issues must be resolved to use the new genetic technology wisely and fairly.

LEARNING OBJECTIVES

After studying this chapter, you should be able to

1. Outline the steps of recombinant DNA technology.
2. Explain how genes are cloned.
3. Describe how DNA recombinant techniques are used to construct genetic maps.
4. Explain how DNA recombinant techniques are used to diagnose and treat human genetic disorders.
5. Describe the undertaking of the human genome project.
6. Outline other applications of advancing DNA technology.
7. Discuss the ethical, social, and legal issues that are developing along with this advancing technology.

CHAPTER OUTLINE

Genetic Manipulation by Selective Breeding

1. Selective breeding is common with domesticated organisms.

Selective Breeding Alters Plants and Animals

1. Selective breeding has been done with dogs. It can be done with microorganisms.

What Is Recombinant DNA Technology?

1. Recombinant DNA technology uses a series of steps to produce clones of inserted DNA segments.

Cloning Genes Is a Multistep Process

Restriction Enzymes Cut DNA at Specific Sites

1. Restriction enzymes cut DNA to produce single-stranded tails.
2. DNA fragments are sealed by DNA ligase into other DNA molecules to form new DNA recombinant molecules.

Vectors Serve as Carriers of Genes for Cloning

1. Plasmids are often the vectors joined to DNA fragments that are cloned.

Steps in the Cloning Process

1. Recombinant DNA molecules can be composed of plasmid DNA and human DNA.
2. These recombinants can be cloned in bacterial cells.
3. The cloned DNA of these recombinants can be recovered from the bacterial cells and used in other ways.

PCR Can Be Used to Amplify DNA

1. Recombinant DNA technology can be used to produce many copies of a gene and DNA fragment.
2. The PCR is used to copy specific DNA fragments.

Applications of Genetic Engineering

1. Recombinant DNA technology is being used in many fields.

Recombinant DNA Techniques Have Revolutionized Genetic Mapping

1. A gene can be mapped by these techniques.
2. About 4000 genes are known to cause genetic disorders from an estimated 50,000 to 100,000.
3. RFLPs are inherited between generations of families by codominance. They are genetic markers on chromosomes for identifying other genes such as the human gene for CF.

4. CFCTR is the nonfunctional protein in CF.
5. The CF gene has been mapped, identified, and cloned.

Prenatal and Presymptomatic Diagnosis of Genetic Disorders

1. Amniocentesis and chorionic villi sampling are prenatal tests.
2. Cells recovered from these tests can be studied for chemical and chromosomal birth defects.
3. Presymptomatic screening can uncover genetic disorders later in life, such as PKD.
4. The use of presymptomatic screening raises ethical questions.

The Human Genome Project Is an International Effort

1. The human genome consists of 50,000 to 100,000 genes
2. The effort has four main steps: high resolution genetic maps for each human chromosome, physical mapping for each chromosome, cloning each chromosome, and sequencing the genome.
3. The results of the project will be used in many fields.
4. The project has raised legal, ethical, and moral questions.

Gene Transfer Technology Has Many Applications

1. Genes can also be transferred between higher organisms.

Proteins Can Be Manufactured by Recombinant DNA Technology

1. Many human diseases result from the inability of a gene to make a specific protein.
2. Vectors can inject productive genes into genetically defective cells to treat some nonhuman diseases.
3. In other cases DNA technology can make a missing protein (i.e., insulin) that is introduced into the organism with the disease. The use of these proteins (i.e., BST) can be controversial.

Defective Genes Can Be Replaced Via Gene Therapy

1. Methods for introducing cloned genes into human cells are under study.
2. Viral vectors are used for gene replacement to treat SCID.

New Plants and Animals Can Be Created by Gene Transfer

1. Genetically engineered strains of corn will be grown this way. Strains of tomatoes have been grown this way.
2. It has been used to transfer genes for the growth hormone in domesticated animals.

Gene Transfer Technology Raises Ethical Questions

1. How will these new techniques be used?
2. There are very few guidelines for this currently.

STUDY HINTS

1. Before studying the chapter in depth, quickly read over the chapter and write down the several important principles that you see. Then study them is depth. This chapter is comparatively short and offers an opportunity for this approach. If you find it successful, adopt it for future chapters.
2. Do you have a unit exam scheduled soon in your biology course? Have you developed an effective approach for studying by now? Consult the section "Developing Your Abilities To Think And Study" after Chapter One in the textbook for more ideas.

CHAPTER REVIEW

Key Terms

Define these key terms from the chapter:

1. vector
2. clone
3. restriction enyzme
4. palindrome
5. DNA ligase
6. plasmid
7. cloning
8. prenatal testing
9. presymptomatic testing

10. genome

Chapter Concepts

1. Is science enough, standing alone, to solve the problems of society? How will the breakthroughs of science be applied?
2. Should animals be used in research to solve human problems? What ethical issues do you think are relevant here?

Practice Test

TRUE/FALSE

___ 1. All current recognized breeds of dogs belong to many different species.

___ 2. The bacterium *Escherichia coli* lives in the human intestine.

___ 3. The function of DNA ligase is to fragment DNA into many different pieces.

___ 4. Bacterial cells can divide as rapidly as every 20 minutes.

___ 5. About 4000 different human genes are known to be associated with genetic disorders.

___ 6. Cystic fibrosis in humans is caused by a dominant gene.

___ 7. CFTCR is an abbreviation for a human birth defect.

___ 8. Amniocentesis is a form of prenatal testing.

___ 9. PKD usually develops in a child within ten years after birth.

___ 10. The genome is a set of genes that belong to an individual.

___ 11. Each human has an estimated 500,000 genes.

___ 12. Composing a genetic map includes placing genes in the correct order on the chromosome.

___ 13. Adult-onset diabetes is a polygenic trait.

___ 14. High blood pressure in humans is not a polygenic trait.

___ 15. Emphysema is a respiratory disorder.

___ 16. Hemophilia is an X-linked disorder caused by a dominant gene.

___ 17. Cellulase is an enzyme that can break down cellulose.

___ 18. Viral vectors can be retroviruses.

___ 19. Gene transfer has not yet been used in agriculture.

___ 20. Genes can be transferred into gametes in addition to body cells.

MULTIPLE CHOICE

21. Many domesticated plants and animals have been genetically manipulated by ____________ breeding.
   a. natural
   b. nonnatural
   c. nonselective
   d. selective
22. In recombinant DNA technology a vector is a
   a. cell membrane with genetic markers.
   b. genetically transformed cell.
   c. self-replicating form of DNA.
   d. virus that reproduces rapidly.
23. A restriction enzyme
   a. cuts both strands of DNA at a specific site.
   b. inhibits the rate of a metabolic reaction.
   c. limits the growth of a cell population.
   d. ties together lose strands of a DNA molecule.
24. An example of a palindrome in DNA is
   a. AAAT.
   b. ATTA.
   c. TTTA.
   d. TTTT.
25. The function of DNA ligase is to
   a. cut DNA strands into fragments.
   b. increase the rate of DNA replication.
   c. seal together DNA fragments.
   d. unwind the DNA double helix.
26. Plasmids are self-replicating
   a. bacterial cells.
   b. circular DNA molecules.
   c. linear RNA strands.
   d. viruses.
27. The members of a clone are
   a. cancer-causing cells.
   b. gametes.
   c. genetically identical.
   d. reproduced sexually.
28. Which of the following statements is not true about the PCR?
   a. It is capable of amplifying DNA.

b. It is simple.
c. It is time-consuming.
d. It is used to copy DNA fragments.

29. The estimated number of human genes is in the range of
a. 1000 to 5000.
b. 10,000 t0 25,000.
c. 50,000 to 100,000.
d. 500,000 to 100,0000.

30. Gel electrophoresis
a. clones cells.
b. duplicates molecules of DNA.
c. inhibits metabolic reactions.
d. separates fragments of DNA.

31. The incidence of cystic fibrosis is about
a. 1 in 100.
b. 1 in 2,000.
c. 1 in 100,000.
d. 1 in 1,000,000.

32. CFTCR is involved in the transport of ___________ ions across cell membranes.
a. calcium
b. chloride
c. iodide
d. magnesium

33. Chorionic villi sampling is a
a. form of postnatal testing.
b. form of prenatal testing.
c. withdrawal of fetal blood.
d. withdrawal of fetal waste products.

34. The defect gene product of sickle red blood cells is
a. alpha globin.
b. alpha glucose.
c. beta globin.
d. beta glucose.

35. PKD is characterized by the development of
a. cysts in the kidney.
b. phenylalaine in the muscles.
c. pigment in the urine.
d. potassium in the urine.

36. The gene for PKD is located on chromosome
a. 16.
b. 18.

c. 20.
d. 21.

37. Which of the following choices is not an objective of the Human Genome Project?
    a. the cloning of each human chromosome
    b. the development of low-resolution human genetic maps
    c. the physical mapping of each human chromosome
    d. the sequencing of the human genome
38. Select the substance used to treat human anemia.
    a. erythropoietin
    b. human insulin
    c. interleukin-2
    d. superoxide dimutase
39. Select the substance used to treat cancer.
    a. cellulase
    b. epidermal growth factor
    c. erythropoietin
    d. interleukin-2
40. BST is used to increase
    a. corn harvests.
    b. egg laying.
    c. milk production.
    d. soybean production.

COMPLETION

41. ____________ are exact copies of DNA segments
42. EcoRI is a catalyst or ____________ .
43. New combinations of DNA are ____________ DNA molecules.
44. The gene for cystic fibrosis is on human chromosome number ____________ .
45. PKD is a ____________ trait, as the homozygous or heterozygous genotype will be expressed.
46. Schizophrenia is a ____________ trait, as many gene pairs cause it.
47. Alpha-1-antitrypsin regulates the expression of ____________ proteins.
48. Interferons are used to treat ____________ and viral infections.
49. Affected individuals with SCID lack a functional ____________ system.
50. Flavr-Savr is a kind of ____________ grown with gene transfer.

ANSWERS

Key Terms

1. A vector is a self-replicating fragment used in DNA recombinant technology. It can also be an agent, such as a virus, to inject genetic material into a host cell.
2. A clone is a copy of a gene or group of genes in a cell.
3. A restriction enzyme cuts DNA strands at a site for introduction into other cells to produce DNA recombinants.
4. A palindrome is a sequence of DNA bases that reads the same in either direction (i.e., ATTA).
5. DNA ligase is an enzyme that seals a DNA fragment into another DNA molecule to make a new DNA molecule and gene.
6. A plasmid is a self-replicating, circular DNA molecule found in a bacterial cell.
7. Cloning is a process making exact genetic copies.
8. Prenatal testing involves techniques before birth to diagnose genetic disorders.
9. Presympomatic testing is done later in life to detect and treat genetic disorders.
10. The genome is the entire set of genes of the organism.

Chapter Concepts

1. Knowledge is not enough to treat the problems in society unless used properly. Science must merge with other disciplines as it is applied in society. For example, will a gene be used to treat a human birth defect or produce, by accident, a new microorganism for which there is no defense by the human immune system? This is an issue beyond mere biology.
2. The answer to this is open-ended. A pertinent question is how more about humans can we learn without using animal species. Are there alternatives such as computer modeling?

Practice Test

1. False	11. False	21. d	31. b	41. clones
2. True	12. True	22. c	32. b	42. enzyme
3. False	13. True	23. a	33. b	43. recombinant
4. True	14. False	24. b	34. c	44. seven
5. True	15. True	25. c	35. a	45. dominant
6. False	16. False	26. b	36. a	46. polygenic
7. False	17. True	27. c	37. b	47. milk
8. True	18. True	28. c	38. a	48. cancer
9. False	19. False	29. c	39. d	49. immune
10. True	20. True	30. d	40. c	50. tomato

☙ Chapter 10 ❧

EVOLUTION: NATURAL SELECTION AND GENETICS

CHAPTER OVERVIEW

This chapter sets forth the mechanism of evolution. How does it occur? The concept of this process has roots in the historical writings from Lamarck, Hutton, Darwin, Wallace, and many others. A population evolves through genetic variation and natural selection. Genetic variations are established by mutations and increased by the influence of sexual reproduction and other factors. Natural selection is the environmental force that acts on that variation. Natural selection is differential reproduction, as population members with adaptive variations have a greater probability to survive, leave offspring, and change the gene pool of the next generation. As the gene pool of a population changes, it evolves. There are several types of selection, such as directional selection.

Natural selection accounts for the formation of new species.
Reproductive isolating barriers cause the genetic divergence of a population to yield new species. The current concept of evolution continues to be altered. For example, it is not always a gradual process.

LEARNING OBJECTIVES

After studying this chapter you should be able to

1. Trace the historical events that led to the formation of a concept of evolution.
2. Outline the steps of the Wallace-Darwin theory of natural selection.
3. Explain the role of genetic variation in evolution.
4. Explain the role of sexual reproduction in evolution.
5. State the effects of genetic drift and other factors on evolution.
6. Describe how natural selection is differential reproduction.

7. Contrast the three kinds of selection.
8. Outline the model for speciation.
9. Cite some studies documenting the different kinds of selection and the process of speciation.
10. State some of the new models that are changing the concept of evolution.

CHAPTER OUTLINE

Evolution as an Idea

1. Several ideas in nineteenth-century Europe established the background for the concept of organic evolution.

The Natural Sciences and the Idea of Evolution

1. Taxonomy categorizes the diversity of life forms.
2. Linnaeus devised the binomial system of nomenclature.
3. Buffon offered the idea that species evolve.
4. Lamarck proposed the idea of acquired characteristics.
5. Darwin and Wallace, along with the growing knowledge of genetics, replaced the proposals of Lamarck.

Geology, the Age of the Earth, and Evolution

1. By geological time Ussher estimated the age of the Earth.
2. Other approaches followed that were much more accurate.
3. The Earth is about 4.6 billion years old.
4. Hutton proposed the concept of uniformitarianism.
5. Lyell refined and restated this idea.
6. The writings of Lyell influenced Wallace and Darwin.

Natural Selection Is the Mechanism of Evolution

1. How do species change over time? There were many unanswered questions.

Darwin's Voyage on the Beagle

1. Darwin's voyage over the world led to the collection of many life forms.
2. Darwin wrote *The Origin of Species* from his observations.

Wallace's Explorations in South America and Indonesia

1. Wallace studied life forms independently of Darwin.
2. Wallace wrote about his findings and galvanized Darwin's writings.

The Wallace-Darwin Theory of Natural Selection

1. Many variations exist among the species of a population.
2. Unchecked organisms tend to reproduce exponentially. The writings of Malthus helped with the understanding of this.

3. Individuals struggle for existence. Some through their phenotypic variations are more likely to reach reproductive age and leave offspring.
4. These individuals leave more offspring than those individuals with variations that are not as adaptive.
5. Over time the heritable changes in the phenotype and genotype of a population occur, leading to the change of the population.

Linking Natural Selection to Genetics

1. A population is a group of individuals associated geographically and sharing in the same gene pool.
2. There is a distribution of phenotypes in the population because of genetic differences in the gene pool.

New Alleles Are Generated by Mutation

1. A mutation is a heritable change in the DNA.
2. Many mutations are harmful.
3. Some mutations may be adaptive and provide the ultimate source of genetic variation.
4. Mutations with other factors increase the genetic variability of a population.

Sex Reshuffles the Gene Pool

1. Meiosis and sexual reproduction increase genetic diversity.

Genetic Variation Can Be Introduced by Drift and Migration

1. In genetic drift, the gene frequencies can change by chance. The founder effect can enhance this effect.
2. Bottlenecks and migration also change gene pool frequencies.

What Is Natural Selection?

1. Fitness is a measure of an organism's ability to survive and reproduce.
2. Natural selection is differential reproduction. Adaptive phenotypes have a greater chance to survive, reproduce, and leave offspring and genes to the next generation.

Natural Selection Acts on Variation Present in the Population

1. There are three kinds of selection. They produce populations that are better adapted to the environment.

Stabilizing Selection Narrows the Range of Phenotypes

2. This favors the combination of genotypes that produce intermediate phenotypes.

Directional Selection Favors an Extreme Phenotype

1. This favors phenotypes at one end of the range of phenotypes.

2. One example of this is the effect of DDT on insect populations.

Disruptive Selection Favors Individuals at Both Extremes of a Phenotype Range

1. Morphs are the two extreme phenotypes.
2. Studies with butterfly populations document this.

Natural Selection and the Formation of Species

1. A divergence in a gene pool can form races or subspecies.

What Is a Species?

1. A species is one or more populations of interbreeding or potentially interbreeding organisms. The members are reproductively isolated from other organisms.
2. Genetic divergence can also produce different species.

Geographical Isolation and Species Formation

1. Reproductive isolating mechanisms start genetic divergence and can produce subspecies and eventual species.
2. Often these mechanisms start as geographical (i.e., mountain range) and become biological (i.e., sterile offspring).

The Pace of Evolutionary Change Is Not Always Slow

1. Ecotypes result from reproductive isolation.
2. Evolution and species formation are not always gradual. Quantum models of speciation state this idea.
3. According to the punctuated equilibrium model, rapid speciation is interrupted with long periods of little or no change.

STUDY HINTS

1. An important part of understanding the biology of populations is vocabulary. Continually test yourself over your understanding of key terms.
2. After studying the chapter from memory try to list the historical events, in proper order, that led to forming a concept of evolution.

CHAPTER REVIEW

Key Terms

Define these key terms from the chapter:

1. taxonomy

2. uniformitarianism

3. population

4. gene pool

5. mutation

6. fitness

7. natural selection

8. morph

9. subspecies

10. genetic divergence

Chapter Concepts

1. How does our knowledge of genetics help our understanding of evolution?
2. Can the idea of "survival of the fittest" be understood? What is it from the biologist's analysis?

Practice Test

TRUE/FALSE

___ 1. The binomial system of nomenclature uses the genus and species names of the organism.

___ 2. Lamarck proposed a theory of mutation and natural selection.

___ 3. Hutton proposed the idea of uniformitarianism.

___ 4. Darwin wrote *The Origin of Species* based on his observation of different species in Canada.

___ 5. Wallace's explorations were in South America and Indonesia.

___ 6. Malthus wrote that populations do not grow exponentially.

___ 7. Wallace and Darwin did not recognize that variations exist in populations.

___ 8. The gene pool is the genetic content of one organism.

___ 9. A mutation is a heritable change in DNA.

___ 10. The founder effect is most common in large, diverse populations.

___ 11. Natural disasters in a population can account for the bottlenecks.

___ 12. Fitness refers mainly to the physical strength of members in a population.

___ 13. A population of birds evolves to preserve the members of intermediate body size. This is an example of stabilizing selection.

___ 14. The effect of DDT on the evolution of affected insect populations is an example of directional selection.

___ 15. Before the industrialization of England the black form of the peppered moth was more common than the speckled form.

___ 16. After the industrialization of England the speckled form of the peppered moth was more common than the black form.

___ 17. Disruptive selection favors the extremes of a phenotypic range.

___ 18. Organisms of different species often interbreed.

___ 19. A zebroid is an example of a hybrid organism.

___ 20. The concept of punctuated equilibrium states continuous, gradual evolution of a population.

MULTIPLE CHOICE

21. Taxonomy involves
    a. cataloging the many diverse forms of life.
    b. conducting genetic crosses.
    c. observing the solar system.
    d. studying cells under the microscope.

22. *Felix panthera* is the binomial name of the
    a. house cat.
    b. jaguar.
    c. lion.
    d. tiger.
23. The estimated age of the Earth is about ___________ billion years.
    a. 4.6
    b. 5.8
    c. 6.6
    d. 8.8
24. Charles Lyell was a
    a. biologist.
    b. chemist.
    c. geologist.
    d. physicist.
25. The *Beagle* was the name of a
    a. book.
    b. country.
    c. river.
    d. ship.
26. Wallace conducted his observations along the ___________ River.
    a. Amazon
    b. Mississippi
    c. Nile
    d. Rio Grande
27. Wallace conducted his studies in the
    a. 1600s.
    b. 1700s.
    c. 1800s.
    d. 1900s.
28. The author of *The Origin of Species* was
    a. Buffon.
    b. Darwin.
    c. Lyell.
    d. Wallace.
29. The idea of the exponential growth of populations originally was stated by
    a. Darwin.
    b. Lyell.
    c. Malthus.
    d. Wallace.

30. The details of the source of genetic variation were first stated from the experiments of
    a. Buffon.
    b. Lyell.
    c. Lamarck.
    d. Mendel.
31. NeoDarwinism is an alliance of
    a. anatomy and physiology.
    b. cytology and histology.
    c. evolution and genetics.
    d. populations and communities.
32. The gene pool relates specifically to the
    a. cell.
    b. organism.
    c. population.
    d. tissue.
33. A DNA base conversion of A to T is an example of a
    a. chromosome.
    b. founder effect.
    c. gene pool.
    d. mutation.
34. Which of the following is not true about a genetic drift of a population?
    a. It experiences a change in gene frequencies.
    b. It is very small.
    c. It is very large.
    d. It undergoes a change in its gene pool.
35. HD directly affects the ___________ system.
    a. circulatory
    b. immune
    c. nervous
    d. skeletal
36. Natural selection is
    a. differential reproduction.
    b. classifying organisms.
    c. eliminating organisms.
    d. random reproduction.
37. The morphs in butterflies refer to
    a. ecological niches.
    b. ecotones.
    c. habitats.
    d. phenotypes.

38. The main criterion to distinguish between species is
    a. body size.
    b. body color.
    c. geographic preference.
    d. reproduction.
39. The use of DDT promoted __________ selection in insect population.
    a. directional
    b. disruptive
    c. reproductive
    d. stabilizing
40. *Bufo americanus* is a kind of
    a. bird.
    b. fish.
    c. snake.
    d. toad.

COMPLETION

41. The binomial system of nomenclature refers to the __________ and species of an organism.
42. Lamarck used the example of the length of giraffe __________ to explain the passage of acquired traits.
43. Lyell wrote the book *Principles of* __________ .
44. The Galapagos Islands are off the coast of __________ America.
45. Darwin proposed the idea of __________ selection.
46. A gene pool includes the frequencies of the different __________ or alternative forms of a gene.
47. Fitness is a measure of an organism's ability to survive and __________ .
48. __________ selection favors genotypes that produce intermediate phenotypes.
49. __________ isolating mechanisms produce different species.
50. The __________ models of speciation propose that slow changes produce different species.

ANSWERS

Key Terms

1. Taxonomy is a means of cataloging the diversity of life forms.
2. Uniformitarianism is a concept stating that present geological changes account for past ones.
3. A population is a group of individuals living in the same geographical area and sharing in a common gene pool.
4. The gene pool is the genetic makeup of the population.
5. A mutation is a heritable change in the DNA.
6. Fitness is a measure of an organism's ability to survive and reproduce.
7. Natural selection is differential selection that leads to evolutionary divergence and new species formation.
8. A morph is one of two distinct phenotypes.
9. Subspecies are differential populations from an original one through genetic divergence.
10. Genetic divergence is a separation of a population gene pool into two increasingly distinct new gene pools.

Chapter Concepts

1. A population cannot change without a source of change. Mutations and sexual reproduction establish the variation for change in a population. Before the principles of genetics were related to the aspects of natural selection, a cohesive model of evolution could not be composed.
2. Survival of the fittest does not mean that the study of mere superficial characteristics determines what is better or more adaptive. Instead, reproductive fitness is the key. Organisms are fit to the degree that they survive, reproduce, and affect the gene pool of the next generation of organisms.

Practice Test

1. True	11. True	21. a	31. c	41. genus
2. False	12. False	22. b	32. c	42. necks
3. True	13. True	23. a	33. d	43. *Geology*
4. False	14. True	24. c	34. c	44. South
5. True	15. False	25. d	35. c	45. natural
6. False	16. False	26. a	36. a	46. alleles
7. False	17. True	27. c	37. d	47. reproduce
8. False	18. False	28. b	38. d	48. stabilizing
9. True	19. True	29. c	39. a	49. reproductive
10. False	20. False	30. d	40. d	50. quantum

Chapter 11

EVOLUTIONARY PROCESSES: ABOVE THE SPECIES LEVEL

CHAPTER OVERVIEW

This chapter offers lines of evidence for evolution and describes the evolution of life forms on the Earth over four geological eras. Lines of evidence include fossil study, comparative anatomy and embryo study, comparative biochemistry/DNA study, biogeography, and observations of artificial selection. Evolution has occurred in several patterns, including adaptive radiation.

Over four geological eras life probably originated over three billion years ago from cells in an anaerobic environment. Heterotrophic organisms were followed by autotrophic organisms that added oxygen to the atmosphere. Many new life forms evolved. Among animals, invertebrates were followed by vertebrates. Seedless vascular plants evolved into gymnosperms followed by angiosperms. Animals and plants coevolved against a backdrop of changing physical conditions on the Earth. This evolutionary process established the diversity of life forms classified into five kingdoms today.

LEARNING OBJECTIVES

After studying this chapter, you should be able to

1. Explain how fossil study has provided evidence for evolution
2. Describe the other lines offering evidence for evolution.
3. Outline the pattern of adaptive radiation.
4. Compare some other patterns of evolution.
5. Describe the early conditions of the Earth as life began.
6. List some major events of the Archean/Proterozoic Era.

7. List, in correct order, some of the major events in each period of the Paleozoic Era.
8. List, in correct order, some of the major events in the Mesozoic Era.
9. List some major events in the Cenozoic Era.
10. Briefly describe the five-kingdom scheme of classifying life forms.

CHAPTER OUTLINE

The Fossil Record as It Applies to Evolution

1. What is a fossil?

Types of Fossils and How Fossils Are Formed

1. A fossil is the remains, trace or impression of a prehistoric organism.
2. Sedimentary rocks are often rich in fossils.
3. Each stratum of this rock has a distinctive group of fossils.

Relationship of Fossils to Biological Time

1. In Darwin's time fossils could not be dated.
2. Today fossils can be dated using the half-life of radioactive elements.
3. Dating some fossils has been done through dendrochronology.
4. Geological study has revealed four distinct eras: Archean/Proterozoic, Paleozoic, Mesozoic, and Cenozoic.

Evolution: What Is the Evidence?

1. Microevolution entails small-scale events of evolution.
2. Macroevolution involves large-scale events.
3. Darwin used induction and deduction to arrive at the idea of evolution.

Comparative Anatomy Can Be Used to Trace Evolutionary Change

1. Structures of related organisms often have many similarities (i.e., legs of horses).
2. The forelimbs of vertebrates are homologous structures. Over time the vertebrate limb has been modified by selection.
3. Analogous structures are the result of morphological convergence.
4. Vestigial structures indicate the presence of genetic information that is no longer expressed.

Comparative Embryology Can Establish Common Ancestors

1. The early embryos of related vertebrates are similar in many traits.
2. This pattern indicates modification through evolutionary descent.

Comparative Biochemistry Provides Molecular Insight into Evolution

1. In an evolutionary tree, species with fewer amino acid differences are placed closer together.
2. A closer similarity in amino acid sequences between species indicates a closer genetic structure and evolutionary relationship.

Comparative DNA Studies Link Organisms by Genetics

1. DNA hybridization studies indicate the degree of similarity in the genetic structure of two species.
2. Closer genetic relatedness indicates closer evolutionary relatedness.

Biogeography Divides the World into Biological Zones

1. The distribution of species worldwide is the result of adaptation and evolution.
2. Geography and the fossil record are used to study biogeographical regions.

Artificial Selection Is Analogous to Natural Selection

1. In artificial selection a breeder chooses the genetic variants to mate.
2. Artificial selection offers insight to evolution.

There Are Patterns in the Evolutionary Process

1. Organisms have evolved by certain patterns.

Adaptive Radiation Fills Empty Habitats by Rapid Speciation

1. Adaptive radiation is the evolution of many different species from one species. The evolution of mammals is one example.
2. Darwin's studies of populations on the Galapagos (i.e., birds) lend evidence to this process.

Extinction Is Caused by Environmental Change

1. An extinct species disappears, often rapidly.
2. Five major extinctions have occurred during evolutionary history.

Convergent and Parallel Evolution Are Responses to Selection

1. Divergent evolution can occur through adaptive radiation.
2. Convergent evolution involves similar structures evolving from distant ancestors. The body form of a fish and dolphin is one example.
3. Parallel evolution is the evolution of similar structures in similar lines of descent (i.e., canines of saber-toothed animals.

We Can Reconstruct the Large-Scale Events in Evolution

1. Biological and physical domains coevolved.
2. Plate tectonics is the backdrop for the evolution of life.

Archean/Proterozoic Era Began with the Formation of the Earth

1. The crust of the Earth (4.6 billion years old) formed about 3.8 billion years ago.
2. The primitive atmosphere lacked oxygen. Therefore, the first organisms were heterotrophic.
3. Photosynthesis evolved later in this era and, therefore, autotrophic organisms also evolved.
4. Photosynthesis added oxygen to the atmosphere and allowed aerobic life forms to evolve.
5. Eukaryotic cells evolved 1.2 to 1.4 billion years ago.
6. Eukaryotic cells may have evolved from a symbiotic relationship with prokaryotic organisms.
7. There is little evidence about the evolutionary transition from unicellular to multicellular eukaryotes.

Paleozoic Era Was a Period of Shifting Landmasses

1. This era began 570 million years ago. It has six geological periods.
2. In the Cambrian period trilobites became extinct. The first vertebrates appeared.
3. In the Ordovician period many invertebrates became extinct. Ostracoderms diversified.
4. In the Silurian period jawed fish evolved and seedless vascular plants evolved.
5. The Devonian period was the Age of Fish. Amphibians evolved and the seedless vascular plants diversified.
6. In the Carboniferous period reptile evolved (amniote egg) and gymnosperms evolved.
7. In the Permian period the largest mass extinction of invertebrates occurred. Gymnosperms became diverse and abundant.

Mesozoic Era Was the Age of the Dinosaurs

1. This era began about 245 million years ago.
2. It had three periods (Triassic, Jurassic, and Cretaceous) and geological upheaval (i.e., breakup of Pangea).
3. Invertebrates diversified and reptiles flourished.
4. The ancestors to endothermic organisms, birds and mammals, were established.
5. Angiosperms began to flourish.
6. The era ended with mass extinctions, including the dinosaurs.

Cenozoic Era Includes the Present Day

1. This era began about 65 million years ago.
2. Continents shifted and new habitats formed.

3. Birds and placental mammals diversified.
4. The first rodents and primates appeared.
5. Angiosperms flourished even more.

The Evolutionary Process and the Five Kingdoms

1. Systematics uses a number of sources to establish a five-kingdom scheme of classification: Monera, Protista, Plantae, Fungi, and Animalae.

STUDY HINTS

1. Continue to preview the chapter before studying it in depth. Use the questions and problems at the end of the chapter as guideposts.
2. Reading some extra literature on the geological history of the Earth can help understand the process of biological evolution. Historical conditions were very different from today. Changes in life forms coevolved with other changes on the Earth.

CHAPTER REVIEW

Key Terms

Define these key terms from the chapter:

1. fossil

2. sedimentary rock

3. half-life

4. microevolution

5. macroevolution

6. homologous

7. analogous

8. vestigial

9. evolutionary tree

10. biogeography

11. artificial selection

12. adaptive radiation

13. anaerobic

14. heterotrophic

15. autotrophic

Chapter Concepts

1. Are you convinced with all of the details of the evolution of life forms on the Earth? Are there gaps in the record? Do you see any inconsistencies?
2. How do the various branches of science relate to each other? How does this chapter show that relationship? As one example, compare chemistry and biology.

Practice Test

TRUE/FALSE

___ 1. Body fossils are the skeletal remains of organisms.

___ 2. The phrase "half-life" refers to part of the lifespan of an organism.

___ 3. Humans currently live in the Cenozoic Era.

___ 4. New species are formed in microevolution.

___ 5. Vestigial structures of an organism remain functional.

___ 6. Some snakes have retained small pelvic and hind limb bones.

___ 7. The study of amino acid structure can be used to compose evolutionary trees.

___ 8. The chimpanzee and gorilla have distinctly different amino acid sequences that indicate their genetic structure.

___ 9. Biogeography is the study of the distribution of life forms across the Earth.

___ 10. In the process of natural selection a breeder selects organisms to mate.

___ 11. Adaptive radiation is the same as convergent evolution.

___ 12. Divergent evolution can produce many different species from one original species.

___ 13. The relatedness of a wallaby and rabbit is one example of the results of convergent evolution.

___ 14. Organisms did not evolve on the Earth until the atmosphere become aerobic.

___ 15. Symbiosis is a process involving only one species.

___ 16. The Devonian period is part of the Mesozoic Era.

___ 17. The amphibians were the first vertebrates to evolve.

___ 18. Angiosperms were the first seed plants to evolve.

___ 19. The body temperatures of ectotherms are determined by the external environment.

___ 20. Most biologists currently recognize five different kingdoms of organisms.

MULTIPLE CHOICE

21. ____________ rocks formed directly from the deposits of sand and mud in river beds.
   a. biological
   b. igneous
   c. metamorphic
   d. sedimentary

22. The daughter isotope of Uranium 238 is
    a. lead 206.
    b. lead 207.
    c. uranium 236.
    d. uranium 237.
23. The __________ period is the oldest period of the Paleozoic Era.
    a. Cambrian
    b. Devonian
    c. Ordovician
    d. Silurian
24. The first amphibians evolved during the __________ period.
    a. Cambrian
    b. Devonian
    c. Permian
    d. Triassic
25. __________ were the first vertebrates to evolve.
    a. amphibians
    b. fish
    c. mammals
    d. reptiles
26. The wings of __________ represent the best example of convergent evolution.
    a. bats and other bats
    b. bats and birds
    c. birds and insects
    d. insects and other insects
27. Homologous structures
    a. are the same as analogous structures.
    b. exist rarely among animals.
    c. have the same arrangement of bones.
    d. lack blood vessels.
28. Cytochrome c is found in the __________ of a cell.
    a. lysosome
    b. mitochondrion
    c. nucleus
    d. ribosome
29. The quagga most resembled a
    a. fish.
    b. horse.
    c. snake.
    d. toad.

30. In biogeography, the world is divided into __________ geographic zones.
    a. two
    b. four
    c. six
    d. eight
31. What kind of birds did Darwin study on the Galapagos Islands?
    a. egrets
    b. finches
    c. robins
    d. sparrows
32. The Koala bear is a(n) __________ mammal.
    a. aquatic
    b. extinct
    c. marsupial
    d. subterranean
33. Select the incorrect association.
    a. bird/penguin
    b. mammal/house mouse
    c. mammal/porpoise
    d. reptile/kowari
34. The crust of the Earth had formed by about __________ billion years ago.
    a. 3.8
    b. 6.0
    c. 6.4
    d. 9.4
35. The Archean Era covers about __________ percent of geological time.
    a. 25
    b. 45
    c. 66
    d. 78
36. Select the animal group not present during the Cambrian Period.
    a. archaeocyathids
    b. brachipods
    c. therapsids
    d. trilobites
37. The __________ egg signified the emergence of reptiles.
    a. amniote
    b. aquatic
    c. calcium
    d. soft

38. The Triassic period was part of the ___________ Era.
    a. Cenozoic
    b. Mesozoic
    c. Paleozoic
    d. Permian
39. The tuataras were a group of
    a. birds.
    b. fish.
    c. mammals.
    d. reptiles.
40. The ___________ are the flowering plants.
    a. algae
    b. angiosperms
    c. ferns
    d. gymnosperms

COMPLETION

41. ___________ are fossilized feces.
42. ___________ is the type of evolution producing species from the species currently inhabiting the Earth.
43. ___________ embryology relates the embryos of different vertebrates to discover evolutionary patterns.
44. The Oriental region is separated from the Palaearctic by the ___________ mountains.
45. The number of major evolutionary extinctions through history is ___________ .
46. Unlike heterotrophic organisms, ___________ organisms can synthesize all of their needed nutrients.
47. ___________ were the common life forms in the oceans 500 to 600 million years ago.
48. The ___________ period was a time of major reef building in the Paleozoic Era.

49. Gymnosperms appeared during the __________ period of the Paleozoic Era.

50. The study of base sequence in the molecule __________ provides evidence for evolutionary relatedness.

ANSWERS

Key Terms

1. A fossil is a remain, trace, or impression of a prehistoric organism.
2. Sedimentary rock develops from solid particles deposited in layers.
3. The half-life of an element is the time required for one-half of an original unstable element to be converted into a stable form of that element.
4. Microevolution involves a small-scale event.
5. Macroevolution involves a large-scale event.
6. Homologous structures have a common evolutionary origin that have become modified somewhat through members of species.
7. Analogous structures have a different origin but similar function.
8. A vestigial structure is a nonfunctional remnant of a once-functional organ.
9. An evolutionary tree diagrams different species for their degree of evolutionary closeness.
10. Biogeography is the study of the distribution of plants and animals over the Earth.
11. In artificial selection, a breeder chooses the mating variants.
12. By adaptive radiation one species diversifies into many species.
13. Anaerobic refers to an environment lacking oxygen.
14. Heterotrophic organisms must obtain many nutrients pre-formed.
15. Autotrophic organisms can make all needed nutrients.

Chapter Concepts

1. There is much evidence that life forms evolved. However, there are many gaps and inconsistencies. How did the dinosaurs become extinct? Where are the transitional forms from unicellular to multicellular eukaryotes?
2. The study of fossil element half-lifes (chemistry) is used to understand evolution (biology) The study of geology is necessary to understand the external environment where life evolved.

Practice Test

1. True
2. False
3. True
4. True
5. False
6. True
7. True
8. False
9. True
10. False
11. False
12. True
13. True
14. False
15. False
16. False
17. False
18. False
19. True
20. True
21. d
22. a
23. a
24. b
25. b
26. c
27. c
28. b
29. b
30. c
31. b
32. c
33. d
34. a
35. b
36. c
37. a
38. b
39. d
40. b
41. coprolites
42. macroevolution
43. comparative
44. Himalayan
45. five
46. autotrophic
47. trilobites
48. Silurian
49. Carboniferous
50. DNA

☙ Chapter 12 ❧

VIRUSES, PROKARYOTES, AND PROTISTS

CHAPTER OVERVIEW

This chapter presents the most current scheme of classifying the diverse array of organisms the inhabit the Earth. Most biologists accept a five-kingdom scheme. Viruses are unique and are not classified into any kingdom. Bacteria belong to the kingdom Monera. The first organisms to evolve are in this kingdom of prokaryotes. Bacteria are important to the ecology and mechanism of human disease on the planet. Protists (Kingdom Protista) are a diverse group. The ones that are more similar to animals are distinguished by their means of locomotion. The algae are more plant-like. The green algae are the evolutionary forebears to the plants of today. The slime molds are a transitional stage between the protists and multicellular organisms.

The other three kingdoms are featured in upcoming chapters: the fungi, plants, and animals.

LEARNING OBJECTIVES

After studying this chapter, you should be able to

1. Outline the most current scheme for classifying organisms.
2. List some distinctive characteristics of viruses.
3. Describe some of the main properties of monerans.
4. Discuss the evolutionary position and significance of the bacteria.
5. Explain how eukaryote organelles may be related to prokaryotes.
6. Describe the characteristics of the various groups of protists.
7. Describe the characteristics of the various groups of algae.
8. Briefly describe the slime molds.

CHAPTER OUTLINE

Species and Classification: The Five Kingdoms

1. At one time organisms were classified as plants or animals.
2. The Monera are the prokaryotes. The Archaebacteria may eventually be placed in a separate kingdom.
3. The Protists are a diverse group of eukaryotes.
4. The fungi absorb and decompose nutrients.
5. Plants make nutrients by photosynthesis.
6. Animals eat other organisms.

Viruses Are Infectious Particles, But Are Not Alive

1. Viruses are submicroscopic particles that can infect. They consist of a protein and a nucleic acid.
2. They are not a member of any one of the five kingdoms.

Viruses Are Intracellular Parasites

1. A bacteriophage is a virus that attacks a bacterial cell, using it to reproduce.
2. Others invade animal cells without killing them.
3. AIDS is caused by a retrovirus.
4. A retrovirus makes DNA from RNA by a reverse transcriptase.

Viruses Cause a Wide Range of Diseases

1. There are many viral human diseases, including measles and chicken pox.
2. Many viruses are very host specific.

Viroids and Prions Are Infective Molecules

1. A viroid lacks a protein coat.
2. A prion is an infective form of a protein.

Monera: The Kingdom of Bacteria

1. It consists of bacteria that are found almost everywhere.
2. The cells are prokaryotic with three general shapes: rod, spiral, and sphere.
3. They can reproduce rapidly by binary fission.
4. Some are autotrophs. Others are heterotrophs.

Archaebacteria Are Among the Most Ancient Organisms on Earth

1. There are three kinds of these bacteria, the oldest organisms known.
2. Methanogens are anaerobes that live in swamps.
3. The thermoacidophiles and halophiles live in extreme habitats.

Most Bacteria Are Classified as Eubacteria

1. Some are chemosynthetic bacteria.

2. Some are photosynthetic, such as the cyanobacteria.
3. Some filamentous bacteria can fix nitrogen.
4. The heterotrophs are a diverse group, including the saprophytes and symbionts.
5. Saprophytes decay matter. Symbionts can be useful for the organisms they live in.

Bacterial Diseases Are a Serious Health Threat

1. Many human diseases are bacterial, such as diphtheria and rheumatic fever.
2. Many of these diseases are now controlled by antibiotics.

The Transition to Eukaryotes Occurred About a Billion Years Ago

1. The details of the evolution of prokaryotes to eukaryotes is unclear.

Some Eukaryotic Organelles May Have Been Prokaryotic Symbionts

1. The first step was mutualism between two prokaryotes.
2. The study of mitochondria and chloroplasts today lend evidence to this idea.

Protists Are a Diverse Group of Eukaryotes

1. Protists were the first eukaryotes to evolve.
2. The diversity among protists may lead to further classification.

Protozoa Are Single-Celled, Motile Organisms

1. They are unicellular heterotrophs.
2. Most use digestive vacuoles.
3. They can reproduce by binary fission. Some can reproduce sexually.

Amoeboid Protozoa Use Pseudopods for Movement

1. They move and feed by pseudopodia.
2. Some marine species have shells, such as the foraminifera.

Many Flagellated Protozoa Cause Disease

1. They move by 9 + 2 flagella.
2. A trypanosome causes African sleeping sickness.

Sporozoans Are Parasitic Protists

1. They lack organelles of locomotion.
2. One example is the species causing toxoplasmosis.
3. The species causing malaria require a mosquito for their life cycle.

Ciliated Protozoans Are Complex Protists

1. Most live in freshwater. Some use trichocysts.
2. They have a macronucleus and micronucleus in each cell.

Algae Evolved Several Times from Protist Ancestors

1. Many are phytoplankton.
2. Eugleinoids are unique phototrophs.
3. The *Euglena* can become heterotrophic. It has an eyespot.
4. Chrysophytes are photosynthetic. Some are diatoms.
5. The dinoflagellates are a third group. Some have bioluminescence.

Red and Brown Algae May Represent Ancient Life-Forms

1. Red algae possess the red pigment phycoerythrin. They are an ancient life form.
2. Most red algae are currently marine.
3. Many red algae build tropical coral reefs.
4. Brown algae have the pigment fucoxanthin.
5. *Fucus* is one member showing the properties of brown algae.
6. Brown algae have an alternation of generations.
7. Brown and red algae both have many commercial values.

Green Algae Have Characteristics of Higher Plants

1. Their body form is highly variable.
2. Higher plants probably evolved from them.
3. They have a specialized kind of sexual reproduction.

Slime Molds Represent a Transition between Protists and Multicellular Organisms

1. Their classification is controversial.
2. They are not closely related to other groups of organisms.

STUDY HINTS

1. After studying the chapter try to rewrite Table 12.1 from memory. Start with a blank page that only has five headings at the top (kingdoms) and the characteristics along the lefthand margin. How well do you know these differences and similarities among the members of the five kingdoms?
2. Continue to answer the questions and problems at the end of each chapter.

CHAPTER REVIEW

Key Terms

Define these key terms from the chapter:

1. intracellular parasite

2. bacteriophage

3. reverse transcriptase

4. binary fission

5. autotroph

6. heterotroph

7. heterocyst

8. saprophyte

9. symbiont

10. antibiotic

11. pseudopodia

12. trichocyst

13. phytoplankton

14. bioluminescent

15. spore

Chapter Concepts

1. What are the advantages of classifying organisms? Do you understand why biologists have adopted this approach? Does it help you understand biology better?
2. Will the current five-kingdom scheme be changed? What changes do you foresee?

Practice Test

TRUE/FALSE

___ 1. The organisms of the Kingdom Monera have a eukaryotic cell structure.

___ 2. A bacteriophage is a virus that invades a bacterium.

___ 3. The HIV is a retrovirus.

___ 4. Shingles is caused by a bacterium.

___ 5. A prion has a complex cell structure.

___ 6. Bacterial cells lack a nuclear membrane in their cells.

___ 7. Bacteria can reproduce by binary fission.

___ 8. Most bacteria are cyanobacteria.

___ 9. A heterocyst is a specialized type of gamete.

___ 10. Rheumatic fever is caused by a virus.

___ 11. Penicillin stops the synthesis of the cell wall in some kinds of bacteria.

___ 12. Eukaryotic cells have many kinds of membrane-bond organelles.

___ 13. Protists have a prokaryotic cell structure.

___ 14. Pseudopodia are the reproductive structures of the amoeba.

___ 15. Sporozoans are parasites.

___ 16. Phytoplankton are photosynthetic organisms.

___ 17. Bioluminescence is a specialized type of photosynthesis.

___ 18. *Sargassum* is a genus of green algae.

___ 19. Green algae have cellulose in their cell wall.

___ 20. Spores can be haploid cells.

MULTIPLE CHOICE

21. Which of the following is not a characteristic of fungi?
    a. It is autotrophic.
    b. It has cell wall present.
    c. It is eukaryotic.
    d. It has mitochondria present.
22. Which kingdom of organisms lacks a cell wall?
    a. Animalia
    b. Monera
    c. Fungi
    d. Plantae
23. Which of the following is not a characteristic of the viruses?
    a. They carry genetic information.
    b. They are classified into a kingdom.
    c. They are infectious particles.
    d. They are submicroscopic.
24. The lysis of a cell means that it
    a. grows.
    b. reproduces asexually.
    c. reproduces sexually.
    d. ruptures.
25. By a reverse transcriptase
    a. DNA directs the synthesis of DNA.
    b. DNA directs the synthesis of RNA.
    c. RNA directs the synthesis of DNA.
    d. RNA directs the synthesis of RNA.
26. Select the disease caused by a virus.
    a. chicken pox
    b. cholera
    c. rheumatic fever
    d. strep throat
27. Which of the following is not a shape of a bacterial cell?
    a. bacillus
    b. coccus
    c. hexagon
    d. spirillum
28. Which of the following is not a type of archaebacteria?
    a. halophile
    b. methanogen
    c. tetragen
    d. thermoacidophile

29. Chlorophyll __________ is contained in the cyanobacteria.
    a. a
    b. b
    c. d
    d. x
30. Most bacteria obtain their energy from the oxidation of
    a. inorganic compounds.
    b. organic compounds.
    c. minerals.
    d. radiant energy.
31. Saporphytes feed on
    a. decaying material.
    b. live plant material.
    c. prey species.
    d. viruses.
32. Select the disease caused by a bacterium.
    a. common cold
    b. gonorrhea
    c. hepatitis A
    d. herpes
33. Antibiotics are produced by
    a. abiotic processes.
    b. animals.
    c. microorganisms.
    d. plants.
34. Eukaryotes probably evolved about __________ billion years ago.
    a. 0.6
    b. 1.2
    c. 2.8
    d. 4.6
35. Trypanosomes move by the action of
    a. cilia.
    b. flagella.
    c. pseudopodia.
    d. spicules.
36. Malaria is caused by a
    a. bacterium.
    b. plant.
    c. protozoan.
    d. virus.

37. Trichocysts serve as a means of
    a. defense.
    b. motility.
    c. reproduction.
    d. sensing.
38. Diatomaceous earth is derived from a(n)
    a. algae.
    b. animal.
    c. fungus.
    d. plant.
39. Fucoxanthin is found mainly in ___________ algae.
    a. brown
    b. green
    c. purple
    d. red
40. The gametophyte and sporophyte are part of the
    a. alternation of generations.
    b. gametogenesis and morphogenesis.
    c. ovulation and sporulation.
    d. unicellular and multicellular cycle.

COMPLETION

41. The action of fungi decomposes ___________ organisms.
42. Viruses are ___________ parasites, as they enter the cells of hosts.
43. A ___________ is the kind of microorganism that causes AIDS.
44. As opposed to autotrophs, ___________ require organic molecules as a source of energy.
45. Cyanobacteria generate the gas ___________ as a product of photosynthesis.
46. Rheumatic fever is caused by a bacterium of the genus ___________ .
47. The amoeba of the genus ___________ lacks mitochondria and has a symbiotic relationship with bacteria.
48. The cells of foraminferans are covered with hard ___________ .
49. The *Euglena* senses light with an ___________ in its cell.
50. The number of known species of green algae is about ___________ .

ANSWERS

Key Terms

1. An intracellular parasite infects a host cell and uses it for its metabolism and reproduction.
2. A bacteriophage is a virus that infects a bacterial cell.
3. Reverse transcriptase is an enzyme that controls RNA to DNA synthesis.
4. Through binary fission a bacterial cell splits into two. This is a means of asexual reproduction.
5. An autotroph makes organic molecules from inorganic substances.
6. A heterotroph require organic molecules ready-made.
7. A heterocyst is a specialized cell of cyanobacteria needed for nitrogen fixation.
8. A saprophyte feeds on dead or decaying matter.
9. A symbiont has a relationship with another organism. It may be harmful (parasite) or beneficial (mutualism).
10. An antibiotic is a substance used to treat a bacterial infection in an organism.
11. Pseudopodia are extensions of the amoeboid cytoplasm for locomotion and feeding.
12. A trichocyst is a barbed, threadlike organelle used by cilia for defense.
13. Phytoplankton are photosynthetic organisms that float in the water.
14. A bioluminescent organism emits a form of light.
15. A spore is a haploid cell for survival in harsh conditions.

Chapter Concepts

1. Learning the properties of an overall group is easier than learning them for each individual species. Science tends to be organized and very systematic. This is one more example.
2. Consider some of the unanswered questions about taxonomy. The protists and monerans are very diverse. The evolutionary bridge from unicellular to multicellular eukaryotes is unclear. The relationship of fungi to the other kingdoms is unresolved.

Practice Test

1. False	11. True	21. a	31. a	41. dead (dying)
2. True	12. True	22. a	32. b	42. intracellular
3. True	13. False	23. b	33. c	43. virus
4. False	14. False	24. d	34. b	44. heterotrophs
5. False	15. True	25. c	35. b	45. oxygen
6. True	16. True	26. a	36. c	46. *Streptococcus*
7. True	17. False	27. c	37. a	47. *Pelomyxa*
8. False	18. Fals	28. c	38. a	48. shells
9. False	19. True	29. a	39. a	49. eyespot
10. False	20. True	30. a	40. a	50. 7000

Chapter 13

DIVERSITY OF FUNGI AND PLANTS

CHAPTER OVERVIEW

This chapter explains many of the successful adaptations of the fungi and plants. Through numerous adaptations these organisms have made successful transitions into new environments from their evolutionary past.

The fungi are a diverse group ranging from the unicellular yeasts to the more complex club fungi. Lichens are symbiotic associations including some kind of fungus.

Plants are multicellular organisms that carry out photosynthesis. Several adaptations promoted their successful transition onto the land from aquatic ancestors. These include the development of root and vascular systems, leaves, flowers, fruits, and seeds. The angiosperms are the dominant land plant today. The dominant sporophyte alternates with a haploid gametophyte in the life cycle.

LEARNING OBJECTIVES

After this chapter, you should be able to

1. Describe some characteristics of the different kinds of fungi.
2. Describe some symbiotic associations involving fungi.
3. Outline the evolution of plants.
4. List the characteristics of the bryophytes.
5. List the characteristics of the ferns.
6. Describe the characteristics of the gymnosperms.
7. Describe the characteristics of the angiosperms.
8. Explain some of the adaptations that have promoted the success of the land plants.

9. Outline the alternations-of-generations life cycle.
10. Discuss the cultural and economic importance of the angiosperms.

CHAPTER OUTLINE

Fungi Are Eukaryotes Unrelated to Plants

1. Fungi made a successful transition onto land independent of the evolution of plants and animals.

Fungi Have Diverse Body Plans and Nutritional Requirements

1. Spores disperse fungi into new habitats.
2. Hyphae are their multicellular/multinucleate filaments. They grow into a mycelium.
3. Fungi are absorptive heterotrophs.
4. Most are saprophytes. Some are parasites
5. The fungi have high commercial value.

Reproduction and Classification of Fungi

1. Fragmentation and spore formation are asexual means.
2. Fusion of hyphae is a sexual means.

Zygomycetes Are Multinucleate Species

1. They have long, multinucleate filaments with haploid spores and sporangia.
2. Sexual reproduction produces zygospores.
3. Haploid hyphae produce spores.

The Sac Fungi (Ascomycetes) Include Yeasts

1. Yeasts (i.e., baker's yeast) are a unicellular form.
2. Other fungi are important in decay and recycling.

Club Fungi (Basidiomycetes) Are a Diverse Group

1. Members include the mushrooms and toadstools.
2. Sexual reproduction forms basidiospores on basidia.
3. Club fungi have economic importance.

Lichens Are Symbiotic Organisms

1. A lichen is an association of a fungi and a photosynthetic unicellular organism. Each organism benefits.
2. A mycorrhiza is an association between a fungus and the roots of a plant.

Plants Are Multicellular, Photosynthetic Organisms

1. They have evolved many adaptations suited for land.

The Evolution and Adaptation of Plants Began with Algal Ancestors

1. Plants made a transition onto land about 500 million years ago from green algae.
2. Leaves are one land-plant adaptation.
3. The life cycle of the land plant is a haploid gametophyte with a diploid sporophyte, the alternation of generations.

The Transition to Land Required Several Adaptations

1. Adaptations include a means of water loss, cuticles and stomata, pollen development, and roots with a vascular system.

Bryophytes Are a Transitional Group of Plants

1. The evolutionary transition is from water to land.
2. They are small with few stomata.
3. Reproduction is sexual and requires water.
4. Rhizoids attach the plant to the substrate.
5. A visible moss is a haploid gametophyte.

Vascular Plants: Conquering the Land

1. There are lower vascular plants and seed plants (gymnosperms and angiosperms).

Vascular Plants Resulted from Several Major Evolutionary Advances

1. The advances are the root-leaf-vascular system axis, a reduction of the gametophyte generation, and the development of seeds.
2. Xylem is a tissue transporting water and nutrients.
3. Phloem is a tissue transporting sugars and other products of photosynthesis.
4. The sporophyte generation has become dominant.

Ferns Are Seedless Vascular Plants with a Dominant Sporophyte

1. Ferns have a horizontal stem and upright true leaves with vascular tissue.
2. The prothallus is the gametophyte.
3. Water is needed for fertilization.
4. The zygote develops in an archegonium.

Gymnosperms Are Seed Plants

1. There are four divisions including the conifers.
2. A conifer tree is diploid sporophyte.
3. The female gametophyte develops in an ovule.
4. A pollen grain fertilizes an egg in the ovule.
5. A seed develops in the ovule and turns into a sporophyte.

Angiosperms Are Flowering Plants

1. They are the dominant form of plant life.
2. Flower development is a major adaptation.
3. Flowers are reproductive structures.
4. Many flower-feeding insects coevolved with angiosperms.
5. The species have cultural and economic importance.

Trends in Plant Evolution

1. The sporophyte became dominant in the life cycle.
2. A root-shoot axis developed with specialized vascular tissues.
3. Sexual reproduction became successful with flowers and fruits.

STUDY HINTS

1. Look at the key terms at the end of the chapter. List them on a sheet of paper before studying the chapter. After studying the chapter, try to define as many of them as possible.
2. Write two headings at the top of a sheet of paper: Fungi and Plantae. Can you compare them for similarities and differences? Criteria for comparison include mode of nutrition, cell structure, life cycle, etc.

CHAPTER REVIEW

Key Terms

Define these key terms from the chapter:

1. hypha
2. mycelium
3. basidium
4. lichen
5. gametophyte
6. sporophyte

7. stomata

8. pollen

9. xylem

10. phloem

11. prothallus

12. ovule

Chapter Concepts

1. Life began in water and made a transition onto land. What do you think is the basis for this sequence?
2. If evolutionary success can be judged by diversity, then the angiosperms are very successful. Support this statement.

Practice Test

TRUE/FALSE

___ 1. Most fungi are saprophytes.

___ 2. There are more than 60,000 species of fungi.

___ 3. *Saccharomyces cerevisiae* is the scientific name for the mushroom.

___ 4. A lichen is a symbiotic association between fungus and a bacterium.

___ 5. The sporophyte is the diploid stage of a plant.

___ 6. The cuticle is a vascular tissue that conducts water in plants.

___ 7. Bryophytes lack a sporophyte generation.

___ 8. Xylem conducts water in plants.

___ 9. The gametophyte is the dominant stage in vascular plants.

___ 10. Ferns are vascular plants that lack seeds.

___ 11. A rhizome is a type of flower.

___ 12. The prothallus is a heart-shaped gametophyte.

___ 13. The gymnosperms are the flowering plants.

___ 14. The ovule is the site of meiosis in a conifer.

___ 15. Ferns are currently the dominant plant form on the Earth.

___ 16. The first known fossils of angiosperms are from the Cretaceous.

___ 17. Nectaries are insect feeding stations.

___ 18. The gametophyte of an angiosperm is located in the root system.

___ 19. The pistil is a female reproductive structure in the flower of an angiosperm.

___ 20. The stamen is a male reproductive structure in the flower of an angiosperm.

MULTIPLE CHOICE

21. Select the incorrect statement about the fungi.
    a. Some are multicellular.
    b. Some are unicellular.
    c. Their cell walls contain chitin.
    d. They are prokaryotes.
22. A mycelium is a mass of
    a. hyphae.
    b. sporangia.
    c. spores.
    d. yeast cells.
23. A zygospore develops from a
    a. bud.
    b. haploid cell.
    c. prothallus.
    d. zygote.
24. Yeasts reproduce asexually by
    a. binary fission.
    b. budding.
    c. fertilization.
    d. mitosis.

25. Club-shaped cells are
    a. ascospores.
    b. basidia.
    c. mycorrhizae.
    d. zygospores.
26. Select the genus of the most poisonous mushroom.
    a. *Agricus*
    b. *Aminita*
    c. *Lentinus*
    d. *Saccharomyces*
27. Select the incorrect statement about lichens.
    a. They are heterotrophic.
    b. They are resistant to drought.
    c. They are resistant to low temperatures.
    d. They are symbiotic organisms.
28. Select the following incorrect statement about organisms in the Kingdom Plantae.
    a. They are eukaryotes.
    b. They lack a cell wall.
    c. They are photosynthetic organisms.
    d. They use sunlight.
29. Which of the following is not an adaptation of flowering plants?
    a. flowers
    b. seeds
    c. sporangia
    d. vascular structures
30. It is the Stomata's function to
    a. anchor the plant.
    b. make seeds.
    c. prevent water loss.
    d. produce pollen.
31. The major function of plant roots is to
    a. absorb nutrients and water.
    b. conduct photosynthesis.
    c. reproduce sexually.
    d. store food.
32. Which of the following choices is not a bryophyte?
    a. fern
    b. hornwort
    c. liverwort
    d. moss

33. Select the incorrect association.
    a. antheridium/motile sperm
    b. archegonium/nonmotile egg
    c. phloem/transport
    d. xylem/spore production
34. Vascular plants evolved in the __________ period.
    a. Devonian
    b. Jurassic
    c. Silurian
    d. Triassic
35. *Psilotum* is a
    a. bryophyte.
    b. fern.
    c. gymnosperm.
    d. protozoan.
36. Select the division of gymnosperms with the largest number of species.
    a. Coniferophyta
    b. Cycadophyte
    c. Ginkophyta
    d. Gnetophyta
37. A conifer tree is a
    a. dominant gametophyte.
    b. dominant sporophyte.
    c. flowering angiosperm.
    d. nonflowering angiosperm.
38. The number of known species of angiosperms is closest to
    a. 100,000.
    b. 150,000.
    c. 230,000.
    d. 550,000.
39. Which of the following is not part of the pistil?
    a. anther
    b. ovary
    c. stigma
    d. style
40. Which of the following choices is not a part of the angiosperm flower?
    a. anther
    b. filament
    c. prothallus
    d. sepal

COMPLETION

41. Wine is produced by the action of ___________ cells, one kind of fungus.
42. The Ascomycetes are the ___________ fungi.
43. The Basidiomycetes are the ___________ fungi.
44. The ___________ is the fungus root.
45. In the alternation of generations, the gametophyte develops from a ___________ spore, as it has one set of chromosomes.
46. ___________ attach bryophytes to a substrate.
47. Xylem conducts ___________ and dissolved nutrients.
48. Phloem conducts dissolved ___________ and other products of photosynthesis.
49. In conifers, the female gametophyte develops in a protective structure called the ___________ .
50. ___________ are the reproductive structures of angiosperms.

ANSWERS

Key Terms

1. A hypha is a multinucleate or multicellular filament in fungi.
2. A mycelium is a group of hyphae.
3. A basidium is a club-shaped cell group forming basidiospores.
4. A lichen is a symbiotic association between a fungus and photosynthetic unicellular organism.
5. The gametophyte is the haploid stage of a plant.
6. The sporophyte is the diploid stage of a plant.
7. Somata are openings on the leaf that control water loss.
8. Pollen is the male gametophyte.
9. Xylem is vascular tissue that conducts water and dissolved nutrients.
10. Phloem is vascular tissue that conducts dissolved sugars and other products of photosynthesis.
11. The prothallus is the heart-shaped gametophyte of a fern.
12. The ovule is the site of gametophyte development in a flowering plant.

Chapter Concepts

1. Water is necessary for life. Water is the most common substance in cells. An aquatic environment supplies this along with the other fundamental substances and environmental conditions necessary for life. Even with a transition onto land, organisms do not lose this dependency on water. Angiosperms have vascular tissue. Amphibians need to return to the water to reproduce. Reptiles have an aquatic environment in their land egg. The human body is about two-thirds water.
2. The angiosperms have established species in a wide variety of habitats. Compare this to the reduced success of gymnosperms and the fewer species of this group.

Practice Test

1. True	14. False	27. a	40. c
2. True	15. False	28. b	41. yeast
3. False	16. True	29. c	42. sac
4. False	17. True	30. c	43. club
5. True	18. False	31. a	44. mycorrhiza
6. False	19. True	32. a	45. haploid
7. False	20. True	33. d	46. rhizoids
8. True	21. d	34. c	47. water
9. False	22. a	35. b	48. sugars
10. True	23. d	36. a	49. ovule
11. False	24. b	37. b	50. flowers
12. True	25. b	38. c	
13. False	26. b	39. a	

☙ Chapter 14 ❧

ANIMAL DIVERSITY

CHAPTER OVERVIEW

This chapter explores the diversity of animal life forms on the Earth. Among this diversity there are basic evolutionary trends in body symmetry, the body cavity, structural specialization, and the degree of complexity of organs and organ systems. These trends are apparent when passing through the invertebrate phyla of the sponges, cnidarians, flatworms, nematodes, mollusks, annelids, arthropods, and echinoderms.

Chordates have several unifying features centering on their skeleton and nervous system. Most are vertebrates: fish, amphibians, reptiles, birds, and mammals. Birds and mammals culminate a vertebrate transition from an aquatic ancestry to a land existence with endothermy, complex organ systems, and complex social interactions.

LEARNING OBJECTIVES

1. List several trends in the pattern of animal evolution.
2. Describe the characteristics of the sponges.
3. Describe the characteristics of the cnidarians.
4. Describe the characteristics of the flatworms.
5. Describe the characteristics of the nematodes.
6. Describe the characteristics of the mollusks.
7. Describe the characteristics of the annelids.
8. Describe the characteristics of the arthropods.
9. Describe the characteristics of the echinoderms.
10. List the several common features of chordates.
11. List the characteristics of the fish, amphibians, reptiles, birds, and mammals.
12. Explain the evolutionary relationship between the groups of vertebrates.

CHAPTER OUTLINE

Animal Body Plans, Organization, and Development

1. They are multicellular heterotrophs.
2. They have evolved many adaptations for a successful transition onto land.
3. Many of these adaptations are tied to their nutrition, such as movement.

Evolutionary Origins and Classification of Animals

1. Unicellular eukaryotes appeared about 1.2 to 1.4 billion years ago.
2. The first known animals were soft-bodied with exoskeletons.
3. Many groups of animals evolved.

Trends in Animal Evolution

1. There is a sac-like or a tube-within-a-tube body plan.
2. An asymmetrical body plan became bilaterally symmetrical.
3. The presence of a coelom resulted.
4. Some animals evolved segmentation.

Sponges: Asymmetry and Cell Specialization

1. They are aquatic, mainly marine.
2. They are asymmetrical and sac-like.
3. Specialized structures are epidermal cells, collar cells, spicules, and amoebocytes.
4. They are filter feeders.
5. They can reproduce asexually by budding.
6. Sexually, they have a ciliated larval stage.

Jellyfish, Corals, and Sea Anemones: Cells Organize into Tissues

1. The life cycle varies from polyp to medusa.
2. A mesoglea is found between the two epidermal layers of the body wall.
3. They have evolved adaptations for feeding, movement, and digestion.
4. The symmetry is radial.
5. Sexually they develop a planula larva.

Flatworms Have Bilateral Symmetry and Cephalization

1. The planarians are free-living with movement and cephalization.
2. Respiratory and circulatory systems are absent in free-living members.
3. There are three distinct tissue layers.
4. The two other classes are the tapeworms and flukes, both parasites. During adaptations most of their organs and tissues have been lost.

Nematodes: A Tube-within-a-Tube Body Plan

1. They are tapered at both ends.

2. The species are found in many different habitats.
3. They are common in the soil.
4. They have a pseudocoelom.

An Evolutionary Transition: Animals with Internal Body Cavities

1. External skeletons appeared, offering many advantages.
2. The skeleton, muscles, and locomotion were refined.

Mollusks: Advanced Organ Systems Appear

1. Major body parts are the visceral mass, mantle, and the head/foot.
2. They have well-developed organ systems.
3. The three major classes are the gastropods, bivalves, and cephalopods (i.e., octopus).

Annelids: Segmented Body Plans

1. Repeated segments of these worms allow for many specializations.
2. They have an enlarged coelom.
3. They are in many environments, including the soil (earthworm).
4. Bristles and body wall muscles enhance movement.
5. The circulatory system is closed.
6. The nervous system is well developed.

Arthropods: Jointed Appendages on Segments

1. There are well over 1,000,000 species.
2. Adaptations include a hard exoskeleton and jointed appendages. Organs systems are well-developed.
3. The life cycle is complex and adaptable.
4. Many have a head, thorax, and abdomen.

Two Arthropod Systems Signal Evolutionary Advances

1. The sense organs and complex nervous system promote complex life cycles and behavior.
2. Some have complex life cycles with a metamorphosis.
3. Crustaceans are one group.
4. Insects are another, very diverse group. They co-evolved with flowering plants.

Echinoderms: The Deuterostome Body Plan

1. They are deuterostomes and not protostomes.
2. They are marine with an adult radial symmetry. Many adaptations (feeding pattern) are related to this.
3. The pre-adult is bilaterally symmetrical.

Chordates: Origin of the Vertebrates

1. Unifying characteristics are the notochord, dorsal and hollow nerve cord, and gill slits at some time in development.

Vertebrates Are Animals with Backbones

1. They are a subphylum of the chordates with a serial arrangement of vertebrae.

Fish: Vertebrates with Jaws

1. The jawless fish are the most primitive vertebrates.
2. The appearance of jaws was a major adaptation.
3. Some fish are cartilaginous (i.e., shark).
4. They are many species of bony fish.
5. Ray-finned fish have a swim bladder.
6. Lobe-finned fish (lungfish and crossopterygians) have muscular fins.
7. The crossopterygians are the ancestor to amphibians.

Amphibians: Vertebrates Move Ashore

1. They evolved mechanisms to adapt to land. One example is their means of respiration.
2. Their fertilization requires water.

Reptiles Reproduce without Water

1. They have a fluid-filled egg.
2. Fertilization is internal.
3. They evolved from amphibians.
4. Three major groups remain today.

Birds Control Internal Body Temperature

1. Early ancestors were bird-like reptiles.
2. Feathers of modern-day species provide insulation and contribute to endothermy.
3. Hollow bones and wings promote flight.
4. Social behavior is complex.
5. Organs and systems are complex.

Mammals Nourish Their Young with Milk

1. A large radiation began about 65 million years ago.
2. Locomotion, larger brains, endothermy, and specialized teeth are major adaptations.
3. Monotremes lay eggs.
4. Marsupials have embryo development in a pouch.
5. The placental mammals are the most diverse group.

STUDY HINTS

1. Begin the chapter by composing a test. Try, for example, a match of the invertebrate phyla and some characteristics. A few days later, after study, test your recall of this information from the chapter.
2. Try the same approach from #1 for the groups of vertebrates.

CHAPTER REVIEW

Key Terms

Define these key terms from the chapter:

1. bilateral symmetry
2. spicule
3. medusa
4. planula
5. cephalization
6. coelom
7. mantle
8. abdomen
9. metamorphosis
10. deuterostome
11. notochord

12. endothermy

Chapter Concepts

1. Many adaptations have been described by this point in your study of biology. What is an adaptation? Think about success and evolution.
2. Do organ systems, structures, and other characteristics of animals evolve independently in a species or is there some kind of relationship?

Practice Test

TRUE/FALSE

___ 1. Animals are autotrophs.

___ 2. Most animal bodies have a tube-within-a-tube plan.

___ 3. The coelom is a body cavity.

___ 4. The spicules are the reproductive cells of sponges.

___ 5. The planula is a motile larva.

___ 6. The planarians are parasitic tapeworms.

___ 7. Mollusks lack well-developed organ systems.

___ 8. Annelids have an open circulatory system.

___ 9. All species of arthropods are aquatic organisms.

___ 10. Metamorphosis is an adaptation of arthropods.

___ 11. Echinoderms are protostomes.

___ 12. Echinoderms live in freshwater environments.

___ 13. There are three vertebrate classes of fish.

___ 14. The sharks are bony fish.

___ 15. The crossopterygians are the probable ancestors of amphibians.

___ 16. Amphibians have made a complete transition from the water to the land.

___ 17. Reptiles evolved from amphibians.

___ 18. *Archeopteryx* was a bird-like reptile.

___ 19. Birds have solid bones as an adaptation for flight.

___ 20. The kangaroo is a monotreme mammal.

MULTIPLE CHOICE

21. A true coelom develops within the
    a. ectoderm.
    b. epiderm.
    c. mesoderm.
    d. endoderm.
22. Which of the following is not correct about sponges?
    a. They are aquatic organisms.
    b. They are bilaterally symmetrical.
    c. They contain specialized cells.
    d. They are sessile as adults.
23. Select the incorrect association.
    a. amoebocytes/transport nutrients
    b. epidermal cells/inner layer
    c. medusa/motile and bell-shaped
    d. polyp/attached and sessile
24. Corals belong to the phylum
    a. Arthropoda.
    b. Cnidaria.
    c. Echinodermata.
    d. Mollusca.
25. Cephalization refers to the concentration of sensory organs in the ___________ end of the body.
    a. anterior
    b. lateral
    c. medial
    d. posterior
26. The tapeworm belongs to the phylum
    a. Annelida.
    b. Arthropoda.
    c. Mollucsa.
    d. Platyhelminthes.
27. Which of the following statements does not describe the species of the phylum Nematoda?
    a. They are absent in the soil.
    b. They have cylindrical bodies.
    c. Most are free-living.
    d. Some inhabit humans.
28. Which of the following is not an adaptation of mollusks?
    a. head/foot
    b. mantle

c. true segmentation
d. visceral mass

29. The term *arthropod* means
    a. jointed-foot.
    b. water-dwelling.
    c. with a shell.
    d. without a backbone.
30. Which of the following is not a characteristic of arthropods?
    a. They have appendages that are jointed.
    b. They have complex nervous system.
    c. They have hard endoskeleton.
    d. Their life cycle that is complex.
31. The largest number of species is in the phylum of
    a. annelids.
    b. arthropods.
    c. mollusks.
    d. nematodes.
32. Each of the following is a body region of an arthropod except the
    a. abdomen.
    b. head.
    c. pseudocoelom.
    d. thorax.
33. The crustaceans belong to the phylum of
    a. annelids.
    b. arthropods.
    c. echinoderms.
    d. sponges.
34. Select the correct characteristic of echinoderms.
    a. Adults are very motile.
    b. Adults have a radial symmetry.
    c. They live in freshwater.
    d. They lack a nervous system.
35. Which of the following is not an adaptation of chordates?
    a. exoskeleton
    b. gill slits
    c. nerve cord
    d. notochord
36. The swim bladder controls the ____________ of a fish.
    a. body size
    b. buoyancy
    c. color
    d. reproductive state

37. There are about ___________ known species of amphibians.
    a. 4000
    b. 7000
    c. 11,000
    d, 15,000
38. The allantois and amnion are parts of the reptile's
    a. body color.
    b. egg.
    c. heat.
    d. skin.
39. Endothermy refers to the control of
    a. blood pH.
    b. blood sugar.
    c. body temperature.
    d. reproductive fitness.
40. Select the monotreme mammal.
    a. echidna
    b. kangaroo
    c. koala
    d. opossum

MATCHING

Match each species or phylum to the correct statement. Each choice is used just once.

___ 41. annelids	a. The starfish is a species in this phylum.
___ 42. arthropods	b. The earthworm is a species in this phylum.
___ 43. cnidarians	c. collar cells and spicules are present.
___ 44. echinoderms	d. These species have a polyp and medusa body form.
___ 45. mollusks	e. Insects are members of this phylum.
___ 46. sponges	f. Gastropods are members of this phylum.

COMPLETION

47. ___________ is a term meaning that a coelom is absent.

48. Sponges are ___________ feeders.

49. Mollusks are not deuterostomes but are ___________ .

50. The ___________ are the largest and most diverse group of mammals.

ANSWERS

Key Terms

1. Bilateral symmetry has a single axis of symmetry producing mirror-like halves in the body.
2. A spicule is a skeletal element in sponges.
3. The medusa is the bell-shaped body form in the cnidarian life cycle, alternating with the polyp.
4. The planula is the larva of the cnidarians.
5. Cephalization is the concentration of sensory tissue in the anterior part of the body.
6. A coelom is a cavity between the body wall and digestive tract.
7. The mantle is a membrane that secretes the shell of a mollusk.
8. The abdomen is the posterior region of the insect body.
9. Metamorphosis is the process of changing from one form to another. Insect development is one example.
10. Deuterostomes develop with the first body opening being the anus.
11. A notochord is dorsal support running the length of the chordate.
12. By endothermy an animal can regulate internal body temperature independently of the external environment.

Chapter Concepts

1. An adaptation is a successful solution that has evolved. It promotes survival. Often it is a structure (i.e., brain) or function (i.e., locomotion). It can be a process such as endothermy.
2. Organ systems evolve together, dependent on each other. The radial symmetry of the starfish co-evolved with other associated organ systems. The complex brain of a mammal determines its potential for behavior.

Practice Test

1. False	11. False	21. c	31. b	41. b
2. True	12. False	22. b	32. c	42. e
3. True	13. True	23. b	33. b	43. d
4. False	14. False	24. b	34. b	44. a
5. True	15. True	25. a	35. a	45. f
6. False	16. False	26. d	36. b	46. c
7. False	17. True	27. a	37. a	47. acoelomate
8. False	18. True	28. c	38. b	48. filter
9. False	19. False	29. a	39. c	49. protostomes
10. True	20. False	30. c	40. a	50. placental

Chapter 15

HUMAN EVOLUTION

CHAPTER OVERVIEW

This chapter describes the series of evolving primates that led to hominid and human evolution. Adaptations, including specialized patterns of movement and vision, accompanied this change. Human-like hominids probably evolved about 4 million years ago and the genus *Homo* about 2 million years ago.

The details of the appearance and migration of *Homo sapiens* remain a source of great debate. There are gaps in the fossil record. Inferences from cellular/molecular studies, however, are shedding new information on the history of human evolution. Worldwide migrations with accompanying changes in culture and society are a major part of the picture.

LEARNING OBJECTIVES

After studying this chapter, you should be able to

1. Describe the specialized adaptations that evolved in primates.
2. State the broad outlines of primate evolution.
3. Describe the appearance of the human-like hominids.
4. Describe the appearance of *Homo*.
5. Describe the origin of *Homo erectus*.
6. Contrast the two major theories for the evolutionary origin of *Homo sapiens*.
7. Compare some of the changes in human culture and migration patterns across the world.
8. List some of the unanswered questions about human evolution.

CHAPTER OUTLINE

Defining Primates and Their Evolutionary Adaptations

1. Mammals evolved from primitive reptiles.
2. Dinosaurs became extinct during the Mesozoic.
3. The major orders of placental mammals evolved during this extinction.

Evolution in Primates Is Linked to Changes in the Environment

1. Primates are advanced through a combination of adaptations.
2. There were cyclic changes in the climate during hominid evolution.

Limbs and Hands Are Highly Developed in Primates

1. Primates evolved a highly-specialized forelimb that could rotate.
2. Primates evolved prehensile movement with an opposable thumb.

Stereoscopic and Color Vision Are Adaptations to Primate Habitats

1. Relocation of the eyes increased depth perception.
2. Primates have cones (color vision and visual acuity) and rods in their eyes.

Upright Posture Was a Major Development in Primate Evolution

1. This posture frees the hands for other tasks.
2. Humans evolved bipedal locomotion.

Primate Jaws and Teeth Are Adaptations to Changing Diets

1. There were changes in jaw size and proportion.
2. Dentition patterns became adaptations to diets.

Stages in Primate Evolution: Origin of Apes and Hominids

1. A divergence formed Old World/New World monkeys and apes/humans.
2. Hominids evolved rapidly in Africa with environmental changes.

The Origins of Hominids Is an Evolutionary Puzzle

1. Ramapiths are probably ancestral to orangutans.
2. A separate hominid line was probably formed about 6 to 8 million years ago.

Human-like Hominids Appeared about Four Million Years Ago

1. The earliest is *Australopithecus afarensis.*
2. They walked upright with ape-like body proportions.
3. They are the probable ancestral stock to several other hominids with overlapping habitats.

The Genus Homo *Appeared about Two Million Years Ago*

1. *Homo,* such as *Homo habilis,* evolved then.
2. Brain size and proportions changed.
3. *Homo erectus* evolved.

Homo erectus *Originated in Africa and Migrated to Europe and Asia*

1. It was different from *Homo habilis* in brain size and other physical features.

Homo erectus *Developed Social and Cultural Systems*

1. *H. erectus* disappeared and was replaced by *H. sapiens.*
2. The details of this transition are incomplete.

The Appearance and Spread of *Homo sapiens*

1. Several methods offer evidence for this.

Two Theories Differ on How and Where Homo sapiens *Originated*

1. There is the out of Africa hypothesis, cited by molecular and cellular evidence.
2. There is the regional continuity hypothesis.
3. The debate is not settled and centers now on chromosome analysis.

Who Were the Neandertals and What Happened to Them?

1. They had distinct skeletal features and used handles on stone tools.

Are Neandertals Our Ancestors or a Separate Evolutionary Line?

1. The two hypotheses continue to compete for the correct explanation.
2. Migrations have been a constant feature of human evolution.

The Transition to Agriculture Is a Cultural Adaptation

1. Agriculture and animal breeding occurred during the Neolithic transition.
2. Genetic maps are used to study this.

Human Migration to the Americas Is a Recent Event

1. There are conflicting theories from a variety of studies.
2. Scientists have identified at least three groups of American Indians, backed by genetic studies.
3. Genetic evidence is continually used to construct new models.

STUDY HINTS

1. Quickly preview the questions and problems at the end of the chapter before studying the chapter in depth. This has been suggested several times. Have you found it effective as part of your study plan?

2. Much of this chapter compares human features with evolving features that became human. How many can you name after studying the chapter? Concentrate on characteristics such as locomotion, vision, skull proportions and brain size.

CHAPTER REVIEW

Key Terms

Define these key terms from the chapter:

1. prehensile

2. opposable

3. cone

4. rod

5. fovea

6. primate

7. hominid

8. mosaic evolution

9. out of Africa

10. regional continuity

Chapter Concepts

1. An important idea is that structure determines function. How is this concept demonstrated in the story of human evolution?
2. The record of human evolution is incomplete. Select one major gap.

Practice Test

TRUE/FALSE

___ 1. The dinosaurs became extinct during the Mesozoic Era.

___ 2. For *Homo sapiens, sapiens* is the species name.

___ 3. If digits of a hand are opposable, the thumb cannot touch the other fingers.

___ 4. The function of the cones of the eye increases if the light in a room becomes dim.

___ 5. Bipedal locomotion is an adaptation in humans.

___ 6. As primates evolved the jaw became longer from front to back.

___ 7. A dentition pattern refers to the teeth.

___ 8. The diet of ancestral primates in the trees was mainly fruit.

___ 9. A divergence of primates took place in Africa about 30 million years ago.

___ 10. Orangutans were ancestral to ramapiths.

___ 11. The earliest human fossils are from about 1.8 million years ago.

___ 12. The skulls of australopithecines had relatively small brains.

___ 13. The australopithecines were ancestral to the later hominids.

___ 14. The earliest species thought to belong to the genus *Homo* inhabited Earth about 8 million years ago.

___ 15. *Homo erectus* originated in Australia.

___ 16. *Homo erectus* probably inhabited Earth over 2 million years ago.

___ 17. All scientists agree about the origin of *Homo sapiens*.

___ 18. Migration of human populations has been a constant feature of human evolution.

___ 19. In the Neolithic transition humans changed from an agricultural culture to a hunter-gatherer culture.

___ 20. All scientists agree when human populations first migrated to North America.

MULTIPLE CHOICE

21. Mammalia is the ___________ of humans.
    a. class
    b. kingdom
    c. order
    d. species
22. The first mammals to evolve most closely resembled a
    a. lemur.
    b. shrew.
    c. tarsier.
    d. whale.
23. Prehensile movement is the ability to
    a. chew foods.
    b. grasp objects.
    c. jump.
    d. run.
24. Select the bone of the forelimb.
    a. femur
    b. humerus
    c. ilium
    d. radius
25. The ___________ of the eye are mainly responsible for color vision and visual acuity.
    a. choroids
    b. cones
    c. rods
    d. scleras
26. Human three-dimensional vision is best described as
    a. monocular.
    b. monoscopic.
    c. stereofoveal.
    d. stereoscopic.
27. Select the bone of the pelvis.
    a. humerus
    b. ischium

c. patella
d. radius

28. The hominids originated in
 a. Africa.
 b. Australia.
 c. South America.
 d. Western Europe.
29. Part of the habitat of the ramapiths was
 a. aquatic ranges.
 b. deserts.
 c. open grasslands.
 d. rocks and sand.
30. Which of the following characteristics does not describe members of *Australopithecus afarensis*?
 a. ape-like body proportions
 b. relatively short arms
 c. some human-like features present
 d. walked upright
31. The term *gracile* means
 a. graceful.
 b. human-like.
 c. large.
 d. slow.
32. For *Homo habilis, Homo* refers to the __________ name.
 a. family
 b. genus
 c. order
 d. species
33. Dentition patterns reveal most about a species
 a. body size.
 b. diet.
 c. nervous system.
 d. social patterns.
34. Which of the following characteristics does not describe *Homo erectus?*
 a. receding chin
 b. absent dentition
 c. height similar to humans
 d. low skull
35. The spread of *H. sapiens* from Africa to Europe and Asia is the __________ hypothesis.
 a. biosphere
 b. DNA

c. out-of-Asia
d. regional continuity

36. Populations in ____________ show the greatest diversity in mitochondrial types.
    a. Africa
    b. Asia
    c. China
    d. India
37. The out-of-Africa hypothesis is based on ____________ inheritance.
    a. maternal
    b. noncytoplasmic
    c. paternal
    d. RNA
38. Which of the following activities is not associated with Neandertals?
    a. built graves
    b. lived in caves
    c. lacked cultural advances
    d. lived in settlements
39. The Neolithic transition began about ____________ years ago.
    a. 10,000
    b. 50,000
    c. 100,000
    d. 500,000
40. mtDNA is found in
    a. messenger and transfer RNA.
    b. mitochondria.
    c. nuclear genes.
    d. plasma membranes.

COMPLETION

41. The order of humans is the ____________ .
42. The ____________ are the bones of the wrist.
43. Humans have ____________ perception, because they can focus on objects at long distances.
44. About 1.5 million years ago primates changed diet from simply fruit to also include ____________ .
45. DNA studies are part of ____________ biology as a branch of biology.

46. The arms bones of the australopithecines were long and __________ with respect to shape.
47. The Neandertals became extinct about 30,000 to __________ years ago.
48. *H. erectus* used handles on stone __________ .
49. The __________ transition to a hunter-gatherer culture probably occurred in the Middle East.
50. Berengia was a __________ bridge.

ANSWERS

Key Terms

1. Prehensile refers to a hand movement that allows grasping.
2. An opposable thumb can touch the tips of the other fingers.
3. The cone is a light receptor of the eye for color vision and visual acuity.
4. The rod is a light receptor of the eye for dim light vision.
5. A fovea has the highest concentration of cones in the eye (retina) for sharp vision and color resolution.
6. Primates are a group (order) of mammals with prehensile movement, opposable thumbs, and other adaptations such as stereoscopic vision. They include the lemur and baboon.
7. The hominids are the primate line leading to human evolution.
8. Mosaic evolution involves subgroups (i.e., primates) evolving at different rates, creating a mosaic pattern of subgroups.
9. The out of Africa hypothesis states that *H. sapiens* evolved in Africa after the exodus of *H. erectus.*
10. The regional continuity hypothesis states that *H. erectus* gradually evolved into *H. sapiens.*

Chapter Concepts

1. Examples in the chapter focus on primate evolution. A concentration of cones allows sharp vision and color discrimination. A change in the arrangement of forearm and hand bones changes locomotion patterns.
2. One choice is the two hypotheses to explain human evolution. There is another gap from the first hominids to *Homo.* Another one exists during the establishment of the Neandertals.

Practice Test

1. True
2. True
3. False
4. False
5. True
6. False
7. True
8. True
9. True
10. False
11. False
12. True
13. True
14. False
15. False
16. True
17. False
18. True
19. False
20. False
21. a
22. b
23. b
24. d
25. b
26. d
27. b
28. a
29. c
30. b
31. a
32. b
33. b
34. b
35. d
36. a
37. a
38. c
39. a
40. b
41. primates
42. carpals
43. depth
44. meat
45. molecular
46. curved
47. 50,000
48. tools
49. Neolithic
50. land

☙ Chapter 16 ❧

PLANT STRUCTURE AND FUNCTION

CHAPTER OVERVIEW

This chapter shows the alliance of structure and function in plants. Structures of the root and shoot have evolved as adaptations that work to promote the survival of the plant organism. The roots anchor the plant and absorb water and minerals. The leaf is the photosynthetic structure. Vascular tissues range continuously through the root and shoot. These structures consists of various specialized tissues.

Plant specialization is refined in the angiosperms, the dominant land plants. These flowering plants are either monocots or dicots.

LEARNING OBJECTIVES

After studying this chapter, you should be able to

1. List some adaptations of the dominant land plants.
2. Compare the monocots and dicots for similarities and differences.
3. Describe the three major plant cell types.
4. Describe the four tissue systems in plants.
5. Describe each of the three plant organ systems.
6. Explain the role of plant meristem tissue.
7. Outline the pattern of secondary growth in woody plants.
8. Evaluate the angiosperms for their biological success on the Earth.

CHAPTER OUTLINE

Form and Function in Plants

1. Plants convert carbon dioxide into food.
2. They can reproduce sexually and asexually.
3. Plants evolved from green algae.
4. Most plants can absorb nutrients and water and transport them in vascular tissue.
5. The dominant vascular land plants are seed plants.

Dividing the Angiosperms: Monocots and Dicots

1. The cotyledon is an embryonic leaf.
2. Differences in monocots and dicots center on the number of cotyledons, leaf veins, stems, flower arrangement, and roots.

The Basic Architectural Plan of Plants

1. Plants consists of roots and shoots.

Plants Have Three Major Cell Types

1. Parenchyma cells carry out photosynthesis, respiration, and store food.
2. Collenchyma cells provide support.
3. Sclerenchyma cells offer support.

Four Tissue Systems in Plants

1. The dermal system epidermis covers the leaves plus the young roots and stems. It forms cork in older plants.
2. The vascular system consists of xylem (conducts water and dissolved minerals) and phloem (carries sugar).
3. The ground system makes up most of the young plant.
4. The meristematic tissue is actively dividing in the tips of stems and roots.

Plants Have Three Main Organ Systems

1. The leaves are the photosynthetic structures of the shoot.
2. The roots anchor the plant and absorb water and nutrients.

Structure, Variability, and Function in Leaves

1. The flat monocot leaf has a sheath.
2. The dicot leaf has a blade and stalk.
3. Leaves are mesophytic, xerophytic, and hydrophytic.
4. Leaves have many modifications.

Internal Structure of Leaves

1. The cuticle of the epidermis prevents water loss.
2. The stomata, with guard cells, of the epidermis control the movement of gases.
3. The internal mesophyll (palisade and spongy) carries out photosynthesis.
4. The vascular system conducts water and minerals to the leaf and sugars from the leaf.

Structure and Function of Plant Stems

1. Leaves attach to the stem at nodes. Internodes are between them.
2. Axillary buds form leaves and flowers.
3. In dicots the vascular tissue is in rings. There is an outer cortex and inner pith.

Roots Provide Support and Storage

1. In dicots the primary root is the tap root.
2. Branches of adventitious roots form a fibrous root system.
3. Some roots are modified.
4. The root epidermis is modified for absorption.
5. The vascular tissue is in a central column.

Meristems and Plant Growth

Primary Growth Elongates the Shoots and Roots

1. The apical meristem is at the top of the shoot.
2. The stem differentiates into nodes with leaves.
3. The root tip forms regions: division (tip), elongation, and maturation.

Secondary Growth Makes Woody Plants Thicker

1. Flowering plants are annuals, biennials, and perennials.
2. Woody plants develop an outer bark and inner wood.
3. Secondary growth occurs on both sides of the vascular cambium (xylem to the inside and phloem to the outside).
4. Wood is aged xylem with heartwood and sapwood.

STUDY HINTS

1. Take a sheet of paper and make two columns headed monocots and dicots. Compare each for similarities and differences.
2. Try to outline the chapter after you study it. Make the headings pertain to structure and function: root absorption, vascular transport, etc. Use your own words.

CHAPTER REVIEW

Key Terms

Define these key terms from the chapter:

1. dicot
2. monocot
3. meristem
4. shoot
5. root
6. xerophytic
7. primary root
8. vascular cylinder
9. region of division
10. wood

Chapter Concepts

1. The term **organ** is often applied more to animals structures such as the heart or stomach. Can you explain, however, how it is also applicable to plant structure?
2. Plants and animals carry out the same basic processes of body function. One example is gas exchange. Can you name others?

Practice Test

TRUE/FALSE

___ 1. Ancient plants made a transition from water onto land.

___ 2. Land plants evolved from green algae.

___ 3. Cattails are dicots.

___ 4. The cotyledon is a specialized flower of a plant.

___ 5. Parenchymal cells carry out photosynthesis.

___ 6. Meristematic tissue is found in the tips of roots and stems.

___ 7. Phloem transports sugars.

___ 8. The root is the main photosynthetic region of the plant.

___ 9. Stomata are specialized root hairs of the root.

___ 10. Cells are closely packed in the spongy mesophyll of the leaf.

___ 11. Veins in the monocot leaf are mostly parallel.

___ 12. Leaves are attached to the stem at internodes.

___ 13. The vascular system is continuous between the leaf and stem.

___ 14. Vascular bundles contain xylem and phloem.

___ 15. Dicots form adventitious roots.

___ 16. The epidermis of roots lacks a cuticle.

___ 17. The apical meristem is located at the tip of the shoot.

___ 18. In the root the region of elongation is between the regions of division and maturation.

___ 19. Annuals complete their life cycle in one year.

___ 20. In the woody stem phloem forms on the inner side of the vascular cambium.

MULTIPLE CHOICE

21. Cones are found in
    a. angiosperms.
    b. bryophytes.
    c. ferns.
    d. gymnosperms.

22. Fruits are found in
    a. angiosperms.
    b. bryophtes.
    c. ferns.
    d. gymnosperms.
23. There are about ___________ different known species of dicots.
    a. 50,000
    b. 75,000
    c. 170,000
    d. 450,000
24. Select the cell type with the thickest cell wall.
    a. collenchyma
    b. epidermis
    c. parenchyma
    d. sclerenchyma
25. A fiber is a type of ___________ cell.
    a. collenchyma
    b. meristematic
    c. parenchyma
    d. sclerenchyma
26. Select the vascular tissue.
    a. epidermis
    b. parenchyma
    c. sclerenchyma
    d. xylem
27. Meristematic tissue is ___________ tissue.
    a. aging
    b. embyronic
    c. nonliving
    d. specialized
28. A ___________ plant lives in low levels of soil and water.
    a. aerophobic
    b. hydrophytic
    c. mesophytic
    d. xerophytic
29. A main function of guard cells is to control
    a. food storage.
    b. gas exchange.
    c. predation.
    d. water absorption.

30. The function of the cuticle is to prevent
    a. anchorage of the plant.
    b. photosynthesis.
    c. starch destruction.
    d. water loss.
31. Stomata are mainly found on the __________ of the leaf.
    a. lower epidermis
    b. internal part
    c. upper epidermis
    d. veins
32. The cuticle is found along the __________ of the leaf.
    a. lower epidermic
    b. internal part
    c. upper epidermis
    d. veins
33. Nodes are the location of __________ along the stem.
    a. leaves
    b. roots
    c. root hairs
    d. terminal bud
34. The innermost part of the dicot stem is the
    a. epidermis.
    b. phloem.
    c. pith.
    d. xylem.
35. The edible part of a potato is an underground
    a. flower.
    b. leaf.
    c. root.
    d. stem.
36. One specialized function of the prop root is
    a. flowering.
    b. reproduction.
    c. support.
    d. water transport.
37. Which of the following choices is not an adaptation of the root epidermis?
    a. fibrils at the cell wall
    b. mutualistic associations with microorganisms
    c. presence of a cuticle
    d. presence of root hairs

38. Most monocots are
    a. nonwoody with primary tissue.
    b. nonwoody without primary tissue.
    c. woody with primary tissue.
    d. woody without primary tissue.
39. Carrots are
    a. annuals.
    b. biennials.
    c. not true plants.
    d. perennials.
40. Wood in a tree trunk is aging
    a. bark.
    b. epidermal tissue.
    c. phloem.
    d. xylem.

COMPLETION

Label each one of the following statements as describing a monocot or a dicot.

41. Flowers usually in multiples of three: ___________ .
42. Vascular bundles arranged in a ring: ___________ .
43. One cotyledon: ___________ .
44. Fibrous root system: ___________ .
45. Two cotyledons: ___________ .
46. Taproot usually present: ___________ .
47. Vascular bundles in complex, random arrangement: ___________ .
48. Floral parts in multiples of four or five: ___________ .
49. Veins usually parallel in the leaf: ___________ .
50. Veins usually branched in the leaf: ___________ .

ANSWERS

Key Terms

1. A dicot is an angiosperm that develops from an embryo with two cotyledons.
2. A monocot is an angiosperm that develops from an embryo with one cotyledon.

3. Meristem is actively dividing tissue in the plant.
4. The shoot is the part of the plant that develops above ground.
5. The root is the part of the plant that develops underground.
6. A xerophytic plant develops in dry, hot conditions.
7. The primary root is the first root formed.
8. The vascular cylinder is a central region of xylem and phloem in the central column of the root.
9. The region is the active, mitotic tip of the root.
10. Wood is aging rings of xylem, usually in the dicot stem.

Chapter Concepts

1. An organ is two or more specialized tissues that work together. A leaf, for example, have several tissues: parenchyma, xylem, phloem, etc. Therefore, it is an organ.
2. In addition to gas exchange, consider internal transport, nutrient procurement, and reproduction. Plants and animals carry out the same basic physiological processes by different evolved structures.

Practice Test

1. True	14. True	27. b	40. d
2. True	15. False	28. d	41. monocot
3. False	16. True	29. b	42. dicot
4. False	17. True	30. d	43. monocot
5. True	18. True	31. a	44. monocot
6. True	19. True	32. c	45. dicot
7. True	20. False	33. a	46. dicot
8. False	21. d	34. c	47. monocot
9. False	22. a	35. d	48. dicot
10. False	23. c	36. c	49. monocot
11. True	24. d	37. c	50. dicot
12. False	25. d	38. a	
13. True	26. d	39. b	

Chapter 17

PLANT NUTRITION, TRANSPORT, AND GAS EXCHANGE

CHAPTER OVERVIEW

This chapter describes the various physiological activities carried out by plants. Plants must fulfill the same needs as animals for survival: procuring nutrients, exchanging gases, and transporting materials. Compared to animals, plants have evolved different solutions to overcoming these potential environmental problems.

Land plants obtain nutrients and water from the soil. They exchange gases mainly through their leaves. They transport substances through vascular tissues. They have also evolved primitive means of movements and complex schemes for flowering.

LEARNING OBJECTIVES

After studying this chapter, you should be able to

1. Contrast macronutrients and micronutrients.
2. Explain how plants obtain water and minerals from the soil.
3. Explain how plants transport substances vertically.
4. Compare the various kinds of plant movements.
5. Contrast the responses of long-day and short-day plants.
6. Outline the mechanism of plant flowering.
7. Describe some plant secondary defense mechanisms.

CHAPTER OUTLINE

Plants Have Different Nutritional Needs than Animals

1. Plants make organic nutrients from carbon dioxide and water while using sunlight.

A Balanced Diet for Plants

1. C, H, and O are essential elements.
2. N, P, and K are primary macronutrients.
3. There are also secondary macronutrients and micronutrients.

Soil and Plant Nutrition

Plants Mine the Soil to Obtain Minerals

1. Their functions include structural building, metabolic roles, enzyme activation, and osmotic balance.

Symbiosis and Mineral Uptake

1. Legumes and symbiotic bacteria conduct nitrogen fixation.
2. Other bacteria form nitrates.

How Plants Take Up Water and Nutrients

Water and Minerals Enter Roots by Two Pathways

1. Transport depends mainly on concentration gradients.
2. Water and dissolved minerals enter by the intracellular route and extracellular route (endodermis, Casparian strip).

Xylem, Water Transport, and Transpiration

1. Xylem cells are tracheids and vessel elements.
2. Water is pulled through the xylem from the roots to the leaves by transpiration.
3. The cohesion-adhesion theory explains water transport.

Guard Cells Regulate Transpiration

1. Guard cells change shape to control the opening and closing of stomata.
2. The stomata control gas exchange.

How Plants Transport and Store Nutrients

Phloem, Sugar, and Translocation

1. The phloem cells are sieve elements (connected by sieve plates), companion cells, and vascular parenchyma.
2. Sugars are translocated from sources to sinks.

Food Moves through the Phloem by a Pressure-Flow Mechanism

1. Phloem develops an osmotic pressure that draws water and produces a pressure for movement.

Plants Respond to Environmental Stimuli

Alterations in Growth Patterns Generate Tropisms

1. A plant bends toward light by a phototropism. It is generated by the hormone auxin.
2. A plant responds to gravity by a geotropism.
3. A thigmotropism is a reaction to touch.

Nastic Movements Are Nondirectional Responses to Stimuli

1. Sleep movements occur by changes in the size of ground parenchymal cells.

Plants Respond to Light and Cold

1. Photoperiodism is the response to the relative amounts of light and dark over 24 hours. It is controlled by phytochrome in part.
2. Short-day plants flower when the nights are relatively long.
3. Long-day plants flower when the nights are relatively short.
4. Pfr has two forms that intercovert based on the length and exposure to darkness.
5. Temperature can affect flowering.

Plant Secondary Compounds: Defense Mechanisms

1. They can attract or repel organisms.
2. Antinutrients protect against insects.

Plant Secondary Compounds Are Used by Humans

1. Some are pesticides.
2. Others have commercial value.

STUDY HINTS

1. Organize your study of this chapter around the themes of how a plant functions: obtains nutrients, transport materials, and so forth.
2. Practice vocabulary for the many key terms in this chapter. For example, can you write, from memory, definitions for the various kinds of nutrients and tropisms?

CHAPTER REVIEW

Key Terms

Define these key terms from the chapter:

1. essential element

2. primary macronutrient

3. endodermis

4. transpiration

5. source

6. phototropism

7. geotropism

8. auxin

9. photoperiodism

10. short-day plant

11. long-day plant

12. phytochrome

Chapter Concepts

1. Do you think that the terms **short-day** and **long-day** are the most accurate way to label plants for photoperiodism response?
2. Remember that science is an ongoing process. What concepts of plant function do you think remain unresolved until more research is conducted?

Practice Test

TRUE/FALSE

___ 1. Essential elements are substances that plants can make independently from the environment.

___ 2. Calcium is a plant macronutrient.

___ 3. Mineral nutrients are acquired by the root system of the plant.

___ 4. Plants use nitrogen for the synthesis of protein.

___ 5. More root hairs decreases the surface area of a root.

___ 6. Plants use ATP to transport substances.

___ 7. Plasmodesmata control the size of stomata for leaf gas exchange.

___ 8. The Casparian strip is between the cells of the endodermis.

___ 9. Tracheids and vessel elements are kinds of phloem cells.

___ 10. At maturity vessel elements are dead and hollow.

___ 11. When the guard cells take up potassium ions, they change their shape to close stomata.

___ 12. Companion cells and sieve elements are kinds of xylem cells.

___ 13. Sugar is a product of photosynthesis.

___ 14. Sugar made in a plant tends to move from a sink to a source.

___ 15. Nastic movements of plants depend on the presence of an external stimulus.

___ 16. Geotropism is a response to light.

___ 17. The winding of vine tendrils is an example of thigmotropism.

___ 18. The poinsettia is a short-day plant.

___ 19. Wheat is a short-day plant.

___ 20. Temperature changes in the environment can affect flowering in plants.

MULTIPLE CHOICE

21. Select the macronutrient in plants.
    a. iron
    b. molybdenum
    c. potassium
    d. zinc

22. Select the micronutrient in plants.
    a. boron
    b. magnesium
    c. phosphorous
    d. sulfur
23. The three-part percentage ratio of nutrients in commercial fertilizers usually does not state the amount of
    a. nitrogen.
    b. phosphorous.
    c. potassium.
    d. zinc.
24. Select the legume.
    a. alfalfa
    b. corn
    c. rose bush
    d. wheat
25. Which of the following choices is not a major role of mineral elements in plants?
    a. activation of enzymes
    b. components of structure
    c. maintenance of osmotic balance
    d. storage of glucose
26. The transport of water in plants depends on
    a. concentration gradients.
    b. epidermal cells dying at a certain rate.
    c. lack of energy expenditure.
    d. pumping the water from a source.
27. The main function of the Casparian strip is to
    a. decrease the rate of photosynthesis.
    b. increase the rate of photosynthesis.
    c. permit the entrance of water and solutes into xylem.
    d. stop the entrance of water and solutes into xylem.
28. Tracheids are
    a. cube-shaped.
    b. flat and square.
    c. long and tapering.
    d. short and column-shaped.
29. Water in the plant stem is mainly ___________ for transpiration.
    a. pulled from above
    b. pulled from below
    c. pushed from above
    d. pushed from below

30. Water molecules stick to each other by __________ bonds.
    a. covalent
    b. hydrogen
    c. ionic
    d. photon
31. An example of adhesion is
    a. potassium ions entering cells.
    b. potassium ions exiting from cells.
    c. water molecules attracting to each other.
    d. water molecules attracting to plant cell walls.
32. The main function of the cuticle on the leaves is to
    a. control water loss.
    b. decrease the rate of photosynthesis.
    c. increase transpiration.
    d. regulate the size of stomata.
33. The main ion that is transported by guard cells to control the size of stomata is
    a. iron.
    b. potassium.
    c. selenium.
    d. zinc.
34. Select the cell that is not a type of phloem cell.
    a. companion cell
    b. sieve element
    c. vascular parenchyma
    d. vessel element
35. Auxin is a(n)
    a. enzyme.
    b. hormone.
    c. specific movement.
    d. stimulus.
36. An example of a source in a plant is where
    a. cofactors are not used.
    b. cofactors are used.
    c. sugar is broken down.
    d. sugar is made.
37. One example of a nastic movement in plants is
    a. auxin movement.
    b. geotropism.
    c. sleep movement.
    d. thigmotropism.

38. Select the short-day plant.
    a. dandelion
    b. onion
    c. spinach
    d. wheat
39. Phytochrome is a plant
    a. covering that absorbs water.
    b. covering that is water repellent.
    c. pigment that senses glucose.
    d. pigment that senses light.
40. Winter wheat is usually planted in the
    a. fall.
    b. spring.
    c. summer.
    d. winter.

COMPLETION

41. In plants, nitrogen, phosphorous, and potassium are the ___________ macronutrients, because they are the major nutrients needed in large quantities.
42. In 5-10-5 fertilizer, the 10% refers to the element ___________ .
43. In plant metabolism, chlorine functions for osmotic and ___________ balance.
44. In plant metabolism, molybdenum functions for ___________ fixation.
45. Some bacteria in the soil form nitrate from ___________ while working with legumes as symbionts.
46. Water passes between the cells of the root by the ___________ route.
47. According to the cohesion- ___________ theory the movement of water by transpiration can be explained.
48. Sieve elements are connected to each other by sieve ___________ .
49. ___________ is the response of plants to the relative amounts of light and dark over a 24-hour period.
50. In phytochrome during the daylight, Pr is converted to the form ___________ .

ANSWERS

Key Terms

1. A essential element is acquired by plants during photosynthesis.
2. A primary macronutrient is needed by the plant in relatively large quantities.
3. The endodermis is a layer of cells outside the vascular cylinder of the root.
4. Transpiration is the loss of evaporating water pulled up through the plant.
5. The source is where a sugar is made in the plant.
6. Phototropism is a reaction of a plant to light.
7. Geotropism is a reaction of a plant to gravity.
8. Auxin is a plant hormone that is transported away from light.
9. Photoperiodism is the reaction of a plant to relative amounts of light and dark over 24 hours.
10. A short-day plant flowers when the nights are relatively long and the days short.
11. A long-day plant flowers when the nights are relatively short and the days long.
12. Phytochrome is a plant pigment the controls photoperiodism.

Chapter Concepts

1. Although the terms emphasize the term "day", research evidence shows the plants are reacting to the length of darkness.
2. The mechanism of plant flowering has many mysteries. Mechanisms of plant transport and movement have as may unanswered questions as answered questions.

Practice Test

1. False	14. False	27. d	40. a
2. True	15. False	28. c	41. primary
3. True	16. False	29. a	42. phosphorous
4. True	17. True	30. b	43. ionic
5. False	18. True	31. d	44. nitrogen
6. True	19. False	32. a	45. ammonia
7. False	20. True	33. b	46. extracellular
8. True	21. c	34. d	47. adhesion
9. False	22. a	35. b	48. plates
10. True	23. d	36. d	49. photoperiodism
11. False	24. a	37. c	50. Pfr
12. False	25. d	38. a	
13. True	26. a	39. d	

☙ Chapter 18 ❧

HOMEOSTASIS

CHAPTER OVERVIEW

This chapter provides numerous examples of the major process of body function, homeostasis. Although there are many disturbances in the external environment, the body can sense these changes and respond to maintain a constant internal environment. Through homeostasis the conditions of the internal environment are kept optimal for cell function.

Examples of homeostasis include the concentration of calcium in the blood and the rate of metabolism in the body. The skin and its derivatives constitute an organ system that works homeostatically. All organ systems contribute, however, and work in concert to achieve optimal body function.

LEARNING OBJECTIVES

After studying this chapter, you should be able to

1. Define the internal environment of the organism.
2. Explain the process of homeostasis.
3. Outline the components of a control system.
4. Describe several examples of negative feedback responses in the body.
5. List the organ systems of the body.
6. Describe briefly a function for each organ system.
7. State some functions of the skin and derivatives.

CHAPTER OUTLINE

Homeostasis Maintains the Internal Environment Shared by All Cells

1. It is an aqueous environment surrounding the cells of an organism.

2. It is maintained for maximum efficiency of cell function.
3. Many organ systems contribute to homeostasis.
4. External factors, such as temperature, can disturb the internal environment.
5. Homeostasis counteracts the influence of these factors.

The Internal Environment Has Two Components

1. It is the extracellular fluid, consisting of the blood plasma and the fluid that bathes the cells.
2. Elements that must be controlled are gases, pH, nutrient and waste product concentration, salts and electrolyte concentration, volume, and pressure.

Cybernetics and Control Systems Used in Animals

1. A sensor detects a stimulus or change.
2. An effector is signaled for an appropriate response.

Homeostatic Systems Have Several Properties

1. The nervous and endocrine systems make major contributions.

Reflexes Are Mediated by the Nervous System

1. The components of the nervous reflex engage in the following order: sensor (receptor for stimulus)—sensory nerve—integration—motor nerve—effector.
2. The effector responds by negative feedback.

The Endocrine System Mediates Extrinsic Control by Hormone Secretion

1. PTH stimulates the bone to release calcium if it is decreasing in the blood.
2. The hypothalamus stimulates the pituitary gland that signals the thyroid gland to release thyroxin. It increases metabolism.

Homeostasis Also Uses Intrinsic Control Systems

1. Local control accompanies some cases of extrinsic control. One example are the local changes from a skeletal muscle contraction.

Negative and Positive Feedback Systems Operate in Homeostasis

1. Negative feedback information reverses the trend of the response. One example is the control of calcium in the blood.
2. Positive feedback causes the sensory input to increase and accelerate the response. One example is the increasing rhythmic contractions during childbirth.

Functional Aspects of Homeostasis

All Major Body Systems Are Involved in Homeostasis

1. Eleven body systems contribute to homeostasis: muscular, skeletal, respiratory, circulatory, immune, digestive, excretory, nervous, endocrine, reproductive, and skin/integument.

The Skin and Its Derivatives Make Up the Integumentary System

1. It is the largest organ of the body.

The Skin Is Composed of Two Layers

1. Specializations of the epidermis (outer layer) include the keratinocytes and melanocytes.
2. The dermis is the inner layer with fibers, capillaries, and nerve endings.

Infoldings of the Epidermis Form Follicles and Glands

1. Hair follicles form hair.
2. Sebaceous glands secrete oil.
3. Eccrine and apocrine glands secrete substances.

Hair and Nails Are Skin Derivatives

1. The hair consists of the root and shaft.
2. The arrector pili muscle contracts to conserve body heat.

The Skin Plays an Important Role In Homeostasis

1. The functions include thermoregulation and osmoregulation.

The Skin Is a Protective Barrier between the Internal and External Environments

1. The epidermis blocks the entry of microorganisms and secretes acids.
2. Melanin blocks ultraviolet light.

Temperature Regulation Is Controlled by Blood Flow in the Skin

1. Receptors in the hypothalamus detect body temperature. This brain region coordinates body temperature control.
2. If body temperature increases, the blood vessels of the dermis dilate. Heat escapes from the skin.
3. If body temperature decreases, the opposite response occurs.

The Skin Helps Regulate Water Balance

1. Water leaves through the skin by evaporation and sweating.

The Skin Is a Primary Organ for Sensory Reception

1. Sensory receptors include Meissner corpuscles (touch) and Pacinian corpuscles (pressure).
2. There are other receptors (i.e., temperature).

The Skin Synthesizes a Number of Important Compounds

1. It makes vitamin D for calcium and phosphorous control.

The Skin Is Selectively Permeable

1. Certain fat-soluble substances enter through the skin.

The Integument Has Multiple Roles in Homeostasis

1. It works with other organ systems.

STUDY HINTS

1. On a sheet of paper list the names of the eleven organ systems on the left side. On the right describe a function for each system.
2. List five situations that you think can threaten your body funtion. Examples could include a change in external temperature or diet. Explain in each case how you think your body would repond through homeostasis.

CHAPTER REVIEW

Key Terms

Define these key terms from the chapter:

1. homeostasis

2. internal environment

3. sensor

4. effector

5. negative feedback control

6. positive feedback control

7. integumentary

8. keratin

9. melanin

10. thermiogenesis

11. Meissner's corpuscle

12. cholecalciferol

Chapter Concepts

1. An important idea that reappears in this chapter is that structure determine function. What examples have you learned about this idea from this chapter?
2. The human body is more that just a collection of individual, independent parts. Do the organs of the body have this relationship? Support this position.

Practice Test

TRUE/FALSE

___ 1. Homeostasis is the maintenance of a stable environment that is external to the organism.

___ 2. Carbon dioxide is a waste product synthesized by cells.

___ 3. The plasma is the cellular part of the blood.

___ 4. Carbon dioxide production by cells makes the intracellular fluid more acidic.

___ 5. A sensor is a structure that makes a response.

___ 6. PTH signals the increase of calcium in the blood.

___ 7. Skeletal muscles in the leg are effectors.

___ 8. Most of the body's homeostatic systems repond on the principle of positive feedback.

___ 9. Oxytocin signals the kidney to control water balance.

___ 10. The reproductive system is controlled by the endocrine system.

___ 11. One of the roles of the integument is reproduction.

___ 12. The mucous membrane is not a part of the skin.

___ 13. The epidermis is the thicker, deeper layer of the skin.

___ 14. The dermis consists of several layer of epithelial tissue.

___ 15. The arrector pili is a specialized gland in the skin.

___ 16. Melanin is a pigment found in the skin.

___ 17. The hypothalamus is a region of the brain that controls body temperature.

___ 18. Thermiogenesis includes the generation of heat of the body.

___ 19. Pacinian corpuscles respond to a change in temperature.

___ 20. The skin is a completely permeable membrane.

MULTIPLE CHOICE

21. *Tetrahymena* is a
    a. ciliated bacterium.
    b. ciliated protozoan.
    c. flagellated bacterium.
    d. flagellated protozoan.
22. Which of the following statements is not true about homeostasis?
    a. It is integrated by the nervous system.
    b. It keeps the internal environment optimal.
    c. Large adjustments are usually made.
    d. It is necessary for cell survival.
23. Water makes up about __________ percent of human body weight.
    a. 20 to 25
    b. 35 to 45
    c. 50 to 60
    d. 75 to 80
24. Sodium functions in the human body mainly as a(n)
    a. acid.
    b. electrolyte.
    c. solvent.
    d. sugar.
25. An effector
    a. detects a stimulus.
    b. is a region of the brain.
    c. is a waste product.
    d. makes a response.
26. In a nervous system, reflex a signal is sent from a(n)
    a. effector to a receptor.
    b. receptor to an effector.
    c. response to a muscle.
    d. response to the brain.

27. Osteoclasts affect the activity of
    a. bone cells.
    b. muscle cells.
    c. nervous cells.
    d. red cells.
28. Thyroxin controls the
    a. balance of water in cells.
    b. concentration of sodium in the blood.
    c. contraction of the uterus.
    d. metabolic rate.
29. The hypothalamus directly signals the ___________ gland.
    a. integrator
    b. pineal
    c. pituitary
    d. renal
30. TSH is secreted by the ___________ gland.
    a. adrenal
    b. pituitary
    c. thyroid
    d. thymus
31. Oxytocin signals the ___________ to contract.
    a. brain
    b. kidney
    c. skeletal muscle
    d. uterus
32. The surface area of the human skin is about ___________ square meters.
    a. 1 to 2
    b. 5 to 10
    c. 15 to 20
    d. 30 to 40
33. Which system mainly paves the way for tissue repair?
    a. immune
    b. muscular
    c. skeletal
    d. urinary
34. Which of the following choices is not a cell type of the epidermis?
    a. basal cell
    b. elastic cell
    c. keratinocyte
    d. melanocyte

35. Another name for the sweat gland is the __________ gland.
    a. apocrine
    b. ceruminous
    c. endocrine
    d. sebaceous
36. The lunula is found on the
    a. brain.
    b. finger.
    c. heart.
    d. scalp.
37. Osmoregulation specifically involves controlling
    a. blood sugar.
    b. body temperature.
    c. muscle contraction.
    d. water balance.
38. As body temperature rises the blood vessels of the skin respond by
    a. changing their location.
    b. constricting.
    c. dilating.
    d. growing by mitosis.
39. Select the wandering white blood cell.
    a. erythrocyte
    b. fibrocyte
    c. macrophage
    d. mast cell
40. In the skin, cholecalciferol is converted to vitamin
    a. A.
    b. B.
    c. C.
    d. D.

MATCHING

Match each description, by letter, to the correct organ system. Each letter is used once.

___ 41. circulatory	a. skin is a part of it
___ 42. digestive	b. acts through hormones
___ 43. endocrine	c. eliminates undigested food
___ 44. integumentary	d. eliminates water and metabolic wastes

___	45. muscular	e. provides support with bones
___	46. nervous	f. obtains and eliminates gases
___	47. reproductive	g. transports oxygen in the blood
___	48. respiratory	h. produces sperm and egg cells
___	49. skeletal	i. controls by sending electrical signals
___	50. urinary	j. its structures contract and pull on bones

ANSWERS

Key Terms

1. Homeostasis is the maintenance of a stable internal environment of the organism.
2. The internal environment is the blood plasma and the fluid that bathes the cells of the organism.
3. A sensor is a structure that detects a change.
4. An effector is a structure that makes a response.
5. Through negative feedback control, a stimulus leads to a response that reverses the direction of the response.
6. Through positive feedback control, a stimulus leads to a response that is increased and accelerated.
7. The integumentary system is the skin and its derivatives.
8. Keratin is a protein made by the epidermis.
9. Melanin is a pigment made in the epidermis.
10. Thermiogenesis is heat generation that raises the metabolic rate of the body.
11. A Meissner's corpuscle is a touch receptor in the skin.
12. Cholecalciferol is the precursor to calcium formed in the skin.

Chapter Concepts

1. A body part (i.e., organ) has a composition that complements its ability to carry out certain specialized functions. The skin, for example, could not control body temperature (function) if it did not have blood vessels (structure) that can carry heat to the surface for escape from the body.
2. The organ systems of the body work in harmony and complement one another. More examples will be presented in upcoming chapters. The blood, as one example, cannot carry oxygen to our cells unless the heart (circulatory system) works with the respiratory system for gas exchange. Look at Figure 18.10 for other examples of organ systems and how they can share functions.

Practice Test

1. False
2. True
3. False
4. True
5. False
6. True
7. True
8. False
9. False
10. True
11. False
12. True
13. False
14. False
15. False
16. True
17. True
18. True
19. False
20. False
21. b
22. c
23. c
24. b
25. d
26. b
27. a
28. d
29. c
30. b
31. d
32. a
33. a
34. b
35. a
36. b
37. d
38. c
39. c
40. d
41. g
42. c
43. b
44. a
45. j
46. i
47. h
48. f
49. e
50. d

☙ Chapter 19 ❧

MUSCULAR/SKELETAL SYSTEMS

CHAPTER OVERVIEW

This chapter shows the mechanisms for movement by the animal body. Most of the examples of the skeleton presented involve the human body. The hydrostatic skeleton and endoskeleton are also described. The human skeleton has an axial and appendicular branch. The bones are constantly growing, being remodeled, and facing possible disorders.

There are three types of vertebrate muscle tissue. Skeletal muscle tissue in the muscles can pull on bones. If bones are connected by movable joints, the attached muscles can contract and produce motion . The source of muscle contraction resides at the molecular level, through the interaction of actin and myosin. There are other accompanying molecular changes. In order to contract the skeletal muscle must be signaled by a motor nerve.

LEARNING OBJECTIVES

After studying this chapter, you should be able to

1. Describe the hydrostatic skeleton.
2. Describe the endoskeleton.
3. Contrast the human axial and appendicular skeletons.
4. Describe the composition of bone tissue.
5. Outline the events of bone growth and remodeling.
6. Compare the three kinds of muscle tissue for similarities and differences.
7. Explain how muscles are arranged in antagonistic pairs.
8. Describe the muscle cell at the molecular level.

9. List, in correct order, the major steps of the sliding theory model of muscle contraction.
10. Describe the dependency of the skeletal muscle function on the nervous system.
11. Describe nonmuscular contractile cells and electric organs.
12. Discuss the relationship of bones and muscle function to other structures/ functions of the body.

CHAPTER OUTLINE

Skeletal Systems Provide Support and Protection

1. They accompanied the evolution of multicellularity.

There Are Three Main Types of Skeletal Systems

1. Animals move by contracting muscles pulling on bones

Hydrostatic Skeletons Use Fluid to Provide Support

1. The development of pressure from fluid-filled chambers that are closed produces movement.
2. The muscular patterns of cnidarians and earthworms are examples.

Exoskeletons Provide Support and Protection

1. They provide a hard, jointed covering.
2. They are reformed by molting.

Endoskeletons Are a Form of Connective Tissue

1. They consist of cartilage and bone in most vertebrates.
2. They enclose and protect vital organs.
3. Cartilage is found in joints and flexible structures.

The Musculoskeletal System Is Multifunctional

1. It contributes to homeostasis.
2. It also produces cells for immunity and regulates the calcium in the blood.

The Human Skeleton Has Two Main Components

The Axial and Appendicular Skeletons

1. The axial skeleton is the skull, vertebral column (33 vertebrae), and rib cage (12 rib pairs and sternum).
2. The appendicular skeleton is the bones of the appendages and the pectoral (scapula and clavicle) and pelvic girdles (2 hipbones).
3. The skull includes a cranium with fontanels at birth.
4. Appendicular bones include the scapula, clavicle, humerus, radius, ulna, carpals, metacarpals, femur, patella, tibia, fibula, tarsals, metatarsals, and phlanges.

Bones Are a Living Tissue

1. Osteocytes are embedded in a matrix of collagen fibers.
2. A long bone has compact bone and spongy bone.
3. The osteocytes of compact bone are arranged around a Haversian canal.
4. Spongy bone has an interlocking lattice of plates.
5. The bone is covered with the periosteum.

Bones Grow Until Adolescence

1. They grow by endochondral ossification.
2. Osteoblasts and osteoclasts have opposing functions.
3. The perichondrium forms cartilage.
4. Bone tissue gradually eliminates cartilage.

Adult Bones Are Constantly Being Remodeled

1. Bone density decreases during osteoporosis.
2. Exercise can affect bone density.

Joints Connect Bones

1. Joints can be immovable, partly-movable, or most movable (synovial).
2. Coverings and synovial fluid permits free movement at the synovial joint.
3. Ligaments connect bones.
4. Tendons attach muscles to bones.
5. Bursae reduce friction between muscles and bones.

Skeletal Disorders Are Common

1. Osteoarthritis is the eroding of cartilage at the ends of bones.
2. Rheumatoid arthritis is the thickening of the synovial membrane.

Skeletal Muscle Systems Work in Pairs

1. The muscles in a pair work antagonistically.
2. The flexion and extension of the forearm is one example.
3. Skeletal muscles are stimulated electrically by the nervous system.

Humans Have Three Types of Muscle

1. Skeletal is attached to bones. Cardiac composes the heart. Smooth composes all other internal organs.
2. They differ by absence or presence of striations, cell shaped and branching pattern, and other characteristics.
3. Muscle fibers of a skeletal muscle are bundled at several levels of connective tissue.

Skeletal Muscle Fibers Have a Distinct Structure

1. Each myofibril in a skeletal muscle cell is a series of sarcomeres.
2. Myosin is the thick, central protein in the sarcomere.
3. Actin is the protein attached to the Z line at each end of the sarcomere.

Muscle Contraction Results from Filament Sliding

1. Myosin heads can attach to the actin filaments and slide them toward the center of the sarcomere.
2. Troponin inhibits the association of actin and myosin.
3. If the muscle is stimulated, Ca ions are released and block the action of troponin. The sarcomere shortens.
4. ATP provides energy.
5. The shortening of the sarcomeres accounts for the contraction of the muscle cells and entire muscle.

The Nervous System Controls Muscle Contraction

1. A motor neuron signals a muscle cell at the neuromuscular junction.
2. The motor neuron sends an electrical signal that releases acetylcholine (chemical signal) at this junction.
3. This excites the muscle cell and leads to Ca release.
4. Muscle fibers (cells) respond by all or none.
5. An entire muscle develops different amounts of force based on how many motor units are recruited.

Nonmuscular Cells Can Contract

1. These cells also have actin and myosin.

Electric Organs Are Modified Muscles

1. Electroplates in fish are one example.

Interaction of Muscular and Skeletal Systems

1. Responses range from a weight lifter's motions to the leg movement of a spider.
2. Levers increase the force and/or velocity of response.

STUDY HINTS

1. Look at the key terms at the end of the chapter in the textbook. How many of these can you define after studying the chapter in depth?
2. This is the one of many chapters you will study that explains body functions. In this case the function is movement. List ten different steps of how a muscle contracts from molecular events. List each step on a separate index card. Mix up the cards. Can you place the events in the correct order?

CHAPTER REVIEW

Key Terms

Define these key terms from the chapter:

1. setae

2. molting

3. endoskeleton

4. axial

5. appendicular

6. endochondral ossification

7. osteoclast

8. osteoblast

9. tendon

10. ligament

11. antagonistic muscles

12. muscle fiber

13. sarcomere

14. neuromuscular junction

15. electroplate

Chapter Concepts

1. Does movement require only one organ system or several? If several, how do they relate?
2. How is skeletal muscle contraction a good example of a structure performing by the collective action of individual parts?

Practice Test

TRUE/FALSE

___ 1. A sea anemone has a hydrostatic skeleton.

___ 2. Molting replaces one endoskeleton with another endoskeleton.

___ 3. The sternum is a bone of the appendicular skeleton.

___ 4. The sacrum is a bone of the axial skeleton.

___ 5. Tarsals are bones of the wrist.

___ 6. The ends of ribs are connected to the sternum by cartilage.

___ 7. The femur is a bone of the arm.

___ 8. Chondrocytes are bone cells.

___ 9. Bones are covered by the perimysium.

___ 10. Osteoblasts produce substances.

___ 11. The bones of the cranium are locked together by immovable joints.

___ 12. The synovial joint allows the greatest amount of motion.

___ 13. Tendons connect bone to bone.

___ 14. Antagonistic muscles produce the same motion.

___ 15. Smooth muscle is unstriated and involuntary.

___ 16. Actin is the thicker filament found in the sarcomere.

___ 17. A sarcomere shortens when the heads on the actin filaments pull the myosin filaments.

___ 18. Calcium is needed for the interaction of actin and myosin in the sarcomere.

___ 19. Acetylcholine is released from the dendrites of neurons.

___ 20. Actin and myosin are found in other kinds of eukaryotic cells in addition to muscle cells.

MULTIPLE CHOICE

21. In the body wall of cnidarians, fibers run
    a. circularly in the middle layer of cells.
    b. circularly in the outer layer of cells.
    c. longitudinally in the middle layer of cells.
    d. longitudinally in the outer layer of cells.
22. Setae are found on the body of
    a. annelids.
    b. arthropods.
    c. echinoderms.
    d. mollusks.
23. Select the bone of the appendicular skeleton.
    a. coccyx
    b. fibula
    c. rib
    d. vertebra
24. Which of the following bones is not a skull bone?
    a. maxilla
    b. temporal
    c. patella
    d. zygomatic
25. The human vertebral column consists of ___________ vertebrae.
    a. 18
    b. 24
    c. 26
    d. 33
26. The ___________ is the shoulder blade.
    a. clavicle
    b. humerus
    c. pelvis
    d. scapula
27. The ___________ is the expanded end of a long bone.
    a. diaphysis
    b. epiphysis
    c. fontanel
    d. Haversian canal

28. Select the incorrect association.
    a. diaphysis/compact bone
    b. epiphysis/spongy bone
    c. pelvic girdle/appendicular
    d. radius/axial
29. Skeletal muscles attach to the ___________ of a bone.
    a. endomere
    b. endosteum
    c. perineurium
    d. periosteum
30. Endochondral ossification refers to
    a. cartilage development.
    b. cartilage disease.
    c. muscle contraction.
    d. nerve degeneration.
31. The joints between human vertebrae are
    a. absent.
    b. completely movable.
    c. immovable.
    d. partly movable.
32. Ligaments connect
    a. bone to bone.
    b. bone to muscle.
    c. muscle to muscle.
    d. muscle to nerve.
33. Which of the following characteristics does not describe bursae?
    a. They consist of compact bone.
    b. They are filled with synovial fluid.
    c. They are lined with a synovial membrane.
    d. They are small sacs.
34. Flexing the forearm means to
    a. bend it.
    b. immobilize it.
    c. straighten it out.
    d. swing it in a circle.
35. Each of the following describes cardiac muscle tissue except
    a. capable of contraction.
    b. found in the heart.
    c. striated.
    d. voluntary.
36. The A band of a sarcomere has
    a. actin only.

b. myosin only.
c. actin and myosin.
d. the Z line.

37. In the sarcomere, the cross bridges are found on the
a. actin.
b. calcium.
c. myosin.
d. Z line.

38. The role of troponin is to
a. block contraction.
b. conduct impulses.
c. serve as a neurotransmitter.
d. store ions.

39. Calcium is released from the ___________ of the muscle cell.
a. ER
b. lysosome
c. nucleus
d. plasma membrane

40. A motor unit contains
a. many muscle fibers and many neurons.
b. many neurons and one muscle fiber.
c. many muscle fibers and one neuron.
d. one muscle fiber and one neuron.

COMPLETION

41. Many arthropods have a(n) ___________ instead of a skeleton on the inside.

42. The skeleton of sharks consists entirely of the tissue ___________ .

43. The ___________ is the part of the skull that houses the brain.

44. The ___________ girdle consists of two hipbones.

45. The medullary cavity of a long bone is filled with ___________ marrow.

46. The ___________ canals in the compact bone form a network with nerves and blood vessels.

47. ___________ are cells that destroy bone cells.

48. In osteoporosis, bone density ___________ .

49. ___________ muscle cells are separated by intercalated disks.

50. The ___________ filament model explains the mechanism of muscle contraction.

ANSWERS

Key Terms

1. Setae are bristles on the surface of the annelid body for movement.
2. Molting is the process of shedding an exoskeleton and replacing it.
3. An endoskeleton is an internal supporting skeleton.
4. The axial skeleton is the skull, vertebral column, and rib cage.
5. The appendicular skeleton is the appendage bones plus the pectoral and pelvic girdles.
6. By endochondral ossification, cartilage is formed as a precursor to bone development.
7. An osteoclast is a cell that removes substances from the central cavity of a long bone.
8. An osteoblast is a cell that lays down the minerals in bones.
9. A tendon attaches a muscle to a bone.
10. A ligament attaches a bone to a bone.
11. Antagonistic muscles have opposite actions for body motion.
12. A muscle fiber is a muscle cell.
13. A sarcomere is a box-like segment that can contract in the muscle cell.
14. The neuromuscular junction is where a motor neurons signals a muscle cell.
15. An electroplate is a group of muscle cells in a fish with electrical activity.

Chapter Concepts

1. Bones, muscles, and nerves work in concert to produce body motion. A muscle must contract to an attached bone with a movable joint. A nerve must signal a muscle for this event.
2. A skeletal muscle contracts on a large scale. However, its shortening is the result of the simultaneous contraction of many smaller, repeating units. These are the sarcomeres.

Practice Test

1. True	11. True	21. d	31. d	41. exoskeleton
2. False	12. True	22. a	32. a	42. cartilage
3. False	13. False	23. b	33. a	43. cranium
4. True	14. False	24. c	34. a	44. pelvic
5. False	15. True	25. d	35. d	45. yellow
6. True	16. False	26. d	36. c	46. Haversian
7. False	17. False	27. b	37. c	47. osteoclasts
8. False	18. True	28. d	38. a	48. decreases
9. False	19. False	29. d	39. a	49. cardiac
10. True	20. True	30. a	40. c	50. sliding

Chapter 20

RESPIRATION

CHAPTER OVERVIEW

This chapter compares the various evolutionary strategies for the problem of gas exchange. A body surface must allow for the movement of gases. In a multicellular organisms, these gases must then be transported. Often part of this transportation is conducted by the circulation of the blood.

The human respiratory system is a series of continuous chambers, a respiratory tract, for the inflow and outflow of air. The alveoli in the lungs are the numerous, microscopic ends to this tract that place oxygen and carbon dioxide in a position to exchange with the blood. The exchange is by diffusion. The blood leaves the lungs to supply body cells with needed oxygen while removing the waste product carbon dioxide.

LEARNING OBJECTIVES

After studying this chapter, you should be able to

1. Explain the importance of diffusion to gas exchange.
2. Compare several successful strategies for gas exchange that have evolved among animals.
3. State the parts of the respiratory tract in the order inhaled air passes through them.
4. Describe the function of the chambers of the respiratory tract.
5. Explain the significance of gas exchange in the alveoli.
6. Outline the steps for inspiration and expiration.
7. Describe how oxygen and carbon dioxide are transported in the blood.
8. Explain how the gaseous needs of cells are met locally.
9. Discuss how respiration is controlled.
10. List some respiratory malfunctions.

CHAPTER OUTLINE

Respiration Involves the Exchange of Gases

1. Large animals carry out external respiration.
2. Oxygen and carbon dioxide are the gases exchanged.

Body Plans and Size Are Related to Respiration

1. Unicellular organisms exchange gases by diffusion.
2. Flattened bodies that are thin and tubular exchange gases easily.

Gas Exchange Occurs across a Respiratory Surface

1. It consists of thin, moist epithelial cells.

Several Methods of Respiration Have Evolved

1. In some simple aquatic animals gases are exchanged as water is moved across the body.

The Body Surface Is a Respiratory Organ in Many Aquatic and Terrestrial Animals

1. Mucus-coated cells cover a network of capillaries.
2. Gases reach the blood and are exchanged by diffusion.
3. The skin is a respiratory organ in the frog.

Gills Increase the Surface Area Available for Respiration

1. A gill is a convoluted outgrowth of blood vessels.
2. It has many adaptations for gas exchange.
3. Oxygen is difficult to obtain in water.
4. Gills obtain oxygen by a countercurrent flow.

Tracheal Systems Deliver Oxygen Directly to Cells

1. They are a series of tubes that carry air.
2. Small tracheoles supply cells.
3. This respiration depends on diffusion and is influenced by temperature.

Lungs Provide an Internal Body Surface for Gas Exchange

1. They are ingrowths of the body wall and connect to the outside by tubes.
2. The vertebrate lung has evolved to increasing complexity.

Human Respiration Uses Principles Found in Other Systems

1. An oxygen-containing medium moves over a moist membrane with blood.
2. Oxygen must be transported and diffused at several sites.

The Human Respiratory System

1. Inhaled air passes through a series of chambers of a tract in the following order: nose—pharynx—larynx—trachea—bronchi—bronchioles—alveoli.

2. The nose has several adaptations to prepare the air.
3. The epiglottis prevents swallowed food in the pharynx from entering the respiratory tract.
4. The larynx has vocal cords.
5. The trachea splits into a right and left bronchus.
6. Gas exchange occurs between the alveoli and the blood in nearby capillaries.
7. Each lung is sealed in a pleural cavity of the thoracic cavity.
8. The diaphragm is at the base of the thoracic cavity.

Breathing and External Respiration Ventilate the Lungs

1. Ventilation is breathing in and out.
2. Inspiration occurs if the diaphragm contracts and moves downward. Muscles of the chest cavity also pull the ribs upward and outward.
3. The cavity enlarges and this reduces the pressure in the pleural cavity.
4. The lungs expand and their pressure drops below the pressure in the atmosphere. Air enters the alveoli of the lungs.
5. A reversal of these steps promotes expiration. It begins with the relaxation of the diaphragm and chest wall muscles.

Diseases Can Obstruct the Airways and Reduce Air Flow

1. The air passageways narrow in asthma.
2. Bronchitis is an inflammatory response.
3. CF is a genetic disorder.

The Alveoli Are Centers of Gas Exchange and Transport

1. A difference in the partial pressures of a gas account for its diffusion across the wall of the alveolus.

Hemoglobin Carries Oxygen to the Tissues

1. Hemoglobin picks up oxygen from the alveoli at the alveoli.
2. Hemoglobin delivers oxygen from the blood to the various cells throughout the body. Cells use the gas for metabolism.

Carbon Dioxide Is Transported in Red Blood Cells

1. This gas diffuses from cells into the blood where it is mainly carried in the bicarbonate ion.
2. In the lung, carbon dioxide leaves the blood and passes to the alveolus for elimination from the body.

Respiration Is Controlled by the Brain

1. The diaphragm and chest wall muscles are skeletal muscles. They require a nerve signal to contract.

2. Respiratory centers in the brain control this signaling.
3. Changes in the levels of oxygen and carbon dioxide in the blood affect the rate of inspiration and expiration.

The Respiratory Control Centers Can Malfunction

1. Breathing stops in apnea.
2. The respiratory center malfunctions in SIDS.

Respiration Control Can Adjust to Changing Conditions

1. The atmospheric pressure decreases with higher altitude.
2. The respiratory center compensates for this change.

STUDY HINTS

1. Study the steps for an inspiration and expiration. List the step for each process, in you own words, on a separate index card. Write at least eight steps for each. For each process, mix up the cards and put the steps in the correct order.
2. At this point in studying the textbook, can you produce your own outline of a chapter? This may be another helpful study aid. List the main topics only on a piece of paper. Then write several supporting statements under each in your own words. This should not replace the outline in the study guide, but serve as a supplement.

CHAPTER REVIEW

Key Terms

Define these key terms from the chapter:

1. external respiration

2. capillary

3. countercurrent flow

4. nares

5. thoracic cavity

6. pleura

7. diffusion

8. bronchitis

9. hemoglobin

10. apnea

Chapter Concepts

1. Do you think that evolution appears in this chapter as a recurring theme of biology? Cite some examples.
2. Will the study of circulation in the next chapter help you better understand gas exchange. Why?

Practice Test

TRUE/FALSE

___ 1. In water oxygen diffuses more slowly than carbon dioxide.

___ 2. Epithelial cells provide the surface for gas exchange in animals.

___ 3. The skin is a respiratory surface in eels.

___ 4. Tracheae conduct blood in arthropods.

___ 5. Oxygen gas diffuses more rapidly in the air than in the water.

___ 6. The lung in the salamander consists of many subdivided sacs.

___ 7. The wall of the human trachea contains rings of cartilage.

___ 8. Cilia in the human respiratory tract beat in a direction that conducts particles into the lungs.

___ 9. The esophagus is part of the respiratory tract.

___ 10. As air is inhaled, it passes from the bronchus into the trachea.

___ 11. As air is exhaled, it passes from the trachea into the larynx.

___ 12. The alveoli are exchange sites for gases.

___ 13. The pleura is a thin sheet of epithelium.

___ 14. The diaphragm is between the abdominal and pelvic cavity.

___ 15. When the diaphragm contracts the pleural cavity becomes larger.

___ 16. The airways in the lungs increase in diameter in asthma.

___ 17. Gases are exchanged in the alveoli by active transport.

___ 18. Hemoglobin tends to bond with oxygen in the lungs.

___ 19. The bicarbonate ion tends to form hydrochloric acid.

___ 20. As the altitude increases the atmospheric pressure decreases.

MULTIPLE CHOICE

21. Which of the following choices is not a body composition that promotes gas exchange?
    a. flattened
    b. stratified
    c. thin
    d. tubular
22. Gas transport is favored across a surface if it is
    a. moist but not thin.
    b. neither thin nor moist.
    c. thin and moist.
    d. thin but not moist.
23. A gill is usually a
    a. convoluted outgrowth with blood vessels.
    b. flat surface lacking blood vessels.
    c. group of muscles and nerves.
    d. twisted series of muscles.
24. Spiracles are ____________ in the respiratory structures of insects.
    a. gills
    b. muscles
    c. nerves
    d. openings
25. In fish, oxygen is obtained from the water by a ____________ flow.
    a. comcurrent
    b. countercurrent
    c. hyperosmotic
    d. hypertonic

26. Lung breathing probably evolved about ___________ million years ago.
    a. 100
    b. 200
    c. 300
    d. 400
27. The first structures of the respiratory tract to receive inhaled air are the
    a. alveoli.
    b. bronchioles.
    c. nares.
    d. tracheae.
28. Which structure is common for the passage of food and air?
    a. epiglottis
    b. esophagus
    c. pharynx
    d. trachea
29. The epiglottis is a flap of tissue over the
    a. esophagus.
    b. larynx.
    c. nares.
    d. palate.
30. The function of the epiglottis is
    a. to change the pitch of the voice.
    b. food storage.
    c. gas exchange for oxygen and carbon dioxide.
    d. to prevent food from entering the respiratory tract.
31. Which type of muscle tissue is found in the wall of the trachea?
    a. cardiac
    b. skeletal
    c. smooth
    d. striated
32. The diaphragm is a
    a. cavity around the lung.
    b. dome-shaped muscle.
    c. respiratory membrane.
    d. tract for airflow.
33. The pleural cavities surround the
    a. brain.
    b. diaphragm.
    c. lungs.
    d. pelvis.

34. Intercostal muscles are attached to the
    a. diaphragm.
    b. heart.
    c. hyoid bone.
    d. ribs.
35. The function of the capillaries is mainly
    a. gas exchange.
    b. glucose production.
    c. red blood cell production.
    d. ventilation.
36. Select the incorrect association.
    a. bronchiole/gas exchange
    b. larynx/voice production
    c. nasal cavity/warms and moistens air
    d. pharynx/air transport
37. At an atmospheric pressure of 760 mm, a gas making up 78% of the atmosphere contributes about ___________ mm of that pressure.
    a. 160
    b. 400
    c. 592
    d. 680
38. The most abundant gas in the atmosphere is
    a. argon.
    b. carbon dioxide.
    c. nitrogen.
    d. oxygen.
39. Carbonic acid breaks down into
    a. carbon dioxide and oxygen.
    b. carbon dioxide and water.
    c. hemoglobin and water.
    d. oxygen and water.
40. When you hold your breath, ___________ in the blood.
    a. carbon dioxide and oxygen both decrease
    b. carbon dioxide and oxygen both increase
    c. carbon dioxide decreases and oxygen increases
    d. carbon dioxide increases and oxygen decreases

COMPLETION

41. ___________ are the smallest kind of blood vessels.

42. The two important gases of metabolism are ___________ and carbon dioxide.

43. The "Adams apple" is part of the __________ .

44. The __________ cavity is the chest cavity.

45. Relaxation of the diaphragm __________ the size of the pleural cavity.

46. Gases are transported in the alveoli by simple __________ .

47. CF is a __________ disorder, as it is inherited.

48. __________ is the pigment that transports oxygen.

49. There are __________ receptors in the lungs that sense the amount of air filling them.

50. As altitude decreases the atmospheric pressure __________ .

ANSWERS

Key Terms

1. External respiration involves breathing plus gas transport and gas exchange at the cellular level.
2. A capillary is a microscopic blood vessel for exchange.
3. A countercurrent flow is an opposing flow of water over gills to blood through vessels.
4. The nares are the nostrils of the nose.
5. The thoracic cavity is the chest cavity.
6. The pleura is a thin sheet of epithelium covering each lung.
7. Diffusion is the movement of gases from a region of higher concentration to a region of lower concentration.
8. Bronchitis is the inflammatory response of the respiratory system to irritants.
9. Hemoglobin is a red pigment of the blood that combines with and transports oxygen.
10. Apena is when breathing stops for longer than 10 seconds.

Chapter Concepts

1. The different types of respiratory systems are evolutionary strategies that have evolved successfully. They are the result of genetic variations preserved by natural selection.
2. The blood is part of the circulatory system. Gas exchange places the blood in a position for transport by the blood to serve the metabolic needs of cells in the body. In Chapter 21 you will learn how the lungs work with the heart and the complete explanation of blood circulation.

Practice Test

1. True
2. True
3. True
4. False
5. True
6. False
7. True
8. False
9. False
10. False
11. True
12. True
13. True
14. False
15. True
16. False
17. False
18. True
19. False
20. True
21. b
22. c
23. a
24. d
25. b
26. d
27. c
28. c
29. b
30. d
31. c
32. b
33. c
34. d
35. a
36. a
37. c
38. c
39. b
40. d
41. capillaries
42. oxygen
43. larynx
44. thoracic
45. decreases
46. diffusion
47. genetic
48. hemoglobin
49. stretch
50. increases

Chapter 21

THE CIRCULATORY SYSTEM

CHAPTER OVERVIEW

This chapter compares the circulatory plans of different animals. Special attention is given to the various vertebrate hearts. Most of the information pertains to the human circulatory system with its three components: heart, blood, and blood vessels. A pump (heart) forces a substance (blood) under pressure through a series of vessels to serve body cells. Along with the circulatory system, the lymphatic system has specialized functions.

Through the circulation of blood, body cells receive the substances needed to carry out their metabolism. It also allows them to lose waste products. The circulatory system is subject to certain malfunctions (i.e., hypertension) and genetic disorders (i.e., hemophilia).

LEARNING OBJECTIVES

After studying this chapter, you should be able to

1. List the three main components of a circulatory system.
2. Outline the evolution of the circulatory system.
3. Compare the various vertebrate hearts for similarities and differences.
4. Describe the structures of their hearts with their functions.
5. Contrast arteries, capillaries, and veins for structure and function.
6. Explain the parts of an ECG.
7. Describe characteristics of blood pressure.
8. Trace the flow of blood through the body.
9. List some functions of the lymphatic system.
10. Describe the components and functions of the blood.

11. Explain how circulatory characteristics, such as blood pressure maintenance, are controlled.
12. Explain the basis for disorders and diseases of this system.

CHAPTER OUTLINE

Circulatory Systems Transport Nutrients, Wastes, and Gases

1. This transport is essential for optimal cell function.

Several Types of Transport Systems Have Evolved in Animals

1. Water is the transport medium in sponges.
2. The components of the circulatory system are the blood,heart, and blood vessels.
3. Insects and some invertebrates have an open circulatory system.
4. Echinoderms and the vertebrates have a closed circulatory system.
5. Humans have a cardiovascular system and a lymphatic circulation with an immune system.

The Circulatory System Has Undergone Evolutionary Adaptations

1. It is a single circuit in fish.
2. In mammals it has two loops, systemic and pulmonary.
3. In many vertebrates, oxygen-depleted blood acquires oxygen from the lungs.

Organization of the Heart and Its Evolutionary Adaptations

1. Many invertebrates have a tubular pump.
2. Fish, amphibians, and reptiles have lymphatic hearts.
3. The vertebrate heart has two auricles and two ventricles.
4. The fish heart pumps blood to the gills for oxygenation.
5. The amphibian heart has two sides and three chambers. Blood from the right auricle (low oxygen) and left auricle (high oxygen) mixes in a single ventricle.
6. The four-chambered heart (birds and mammals) has a separate right and left side. Oxygenated blood traveling through the left atrium and ventricle is received from the lungs.
7. In the four-chambered heart, the left side blood with high oxygen does not mix with the low-oxygen blood traveling through the right atrium and ventricle.

The Human Heart Has Two Sides and Four Chambers

1. There are also two pairs of valves: AV and semilunar.

2. The valves prevent the blood from flowing backwards.
3. On each heart side the blood flows from the atrium to the ventricle and out through the ventricle.

There Are Three General Types of Blood Vessels

1. They are arteries, capillaries, and veins.

Arteries Have Thick, Flexible Walls

1. Arteries branch into smaller arterioles.
2. Arteries and arterioles conduct blood away from the heart.
3. Of the three layers in their walls, the middle layer is smooth muscle.
4. The smooth muscle controls blood pressure by expanding and constricting blood vessel size. If the muscle layer contracts, the vessel constricts and reduces blood flow.

Capillaries Are One Cell Thick

1. Arterioles deliver blood to the capillaries.
2. A capillary consists of one layer of endothelial cells.
3. These vessels occur in capillary beds.
4. They exchange substances with body cells.

Veins Operate at Lower Pressure and Are Thinner Than Arteries

1. Capillaries send blood into venules and veins, which return blood to the heart.
2. Valves and skeletal muscle contractions help venous blood return to the heart.

Blood Flows through The Body In a Single Direction

1. The circuits are the pulmonary and systemic.

Heart Muscle Pumps Blood through the Body

1. The cardiac cycle is one heartbeat.
2. When the atria contract, blood enters the ventricles. The AV valves open.
3. When the ventricles contract (systole), the AV valves close and the semilunar valves open. Blood leaves the right ventricle through the pulmonary artery and left ventricle through the aorta.
4. The heart sounds are from the closure of valves.
5. The cardiac cycle occurs at a certain rate.

A Pacemaker Regulates the Heartbeat

1. The SA node stimulates the atria to contract.
2. From the AV node an impulse travels through the bundle of His and Purkinje fibers. The ventricles are stimulated to contract.

Electrocardiograms Record the Electrical Activity of the Heart

1. The P wave represents the conduction of the contraction impulse for atrial systole.
2. The QRS complex represents the conduction of the contraction impulse for ventricular systole.
3. The T wave represents the electrical changes during ventricular relaxation.

Heart Attacks Result from the Death of Cardiac Muscle

1. Blockage of coronary arteries decreases the delivery of blood to the heart muscle.
2. Angina pectoris is gradual and temporary.
3. More severe blockages can lead to the heart attack.

Blood Flow Exerts Pressure on Blood Vessels

1. The origin of the force is ventricular contraction.
2. It can be measured: i.e., 120/80.
3. It varies in animals.
4. It decreases in the organisms as the blood travels further from the heart.
5. The medulla regulates blood pressure.
6. Hypertension is high blood pressure.

The Lymphatic System Is a Secondary Circulatory System

1. Lymph is altered interstitial fluid.
2. It is collected by lymph capillaries.
3. Progressivly larger lymph vessels return the lymph toward the heart.
4. The lymph joins the blood in vessels near the heart.
5. Lymph organs and nodes are part of the system. They have immune functions.

Blood Is a Tissue Composed of Fluid and Cells

1. The blood consists of cells (40% to 50%) and plasma (50% to 60%).

Red Blood Cells Transport Oxygen

1. They are numerous and are packages of hemoglobin at maturity.
2. Bright red blood is due to highly-oxygenated hemoglobin.
3. Hemoglobin is recycled in the body.
4. Sickle cell anemia is a genetic disorder.

White Blood Cells Are Part of the Body's Defense System

1. There are five kinds: neutrophil, eosinophil, basophil, monocyte, and lymphocyte. Each has specific functions.
2. Some are phagocytic.

Platelets Help Form Blood Clots

1. They are formed from megakaryocytes.
2. They start blood clotting.
3. Hemophilia is a genetic disorder.

STUDY HINTS

1. Compose several tables that compare: arteries to veins, two-chambered heart/four-chambered heart, pulmonary and systemic circuits, etc. Compare these and others for similarities and differences.
2. In anticipation of an upcoming test, try to compose some test questions for study. Examples can conclude matches of heart structures and cells of the blood.

CHAPTER REVIEW

Key Terms

Define these key terms from the chapter:

1. closed circulatory system

2. ventricle

3. pulmonary circuit

4. semilunar valve

5. artery

6. capillary

7. vein

8. cardiac cycle

9. sinoatrial node

10. coronary artery

11. hypertension

12. erythrocyte

13. leukocyte

14. platelet

15. hemophila

Chapter Concepts

1. The key concept in this chapter is internal transport. How is it essential for life?
2. Homeostasis is a unifying theme of body function or physiology. How does the circulatory system contribute to homeostasis?

Practice Test

TRUE/FALSE

___ 1. Insects have an open circulatory system.

___ 2. Echinoderms have a closed circulatory system.

___ 3. Fish have a three-chambered heart.

___ 4. Mammals have a four-chambered heart.

___ 5. The AV valves in the heart prevent the blood from flowing backwards into the ventricles.

___ 6. Capillaries are the only kind of blood vessel that exchange materials with cells.

___ 7. The wall of veins has five layers.

___ 8. The contraction of the smooth muscle layer in an arteriole increases the diameter of this blood vessel.

___ 9. As blood circulates the smaller venules combine to form veins.

___ 10. Blood flows from the pulmonary veins into the left atrium.

___ 11. The force of ventricular systole closes the semilunar valves.

___ 12. The SA node is located in the left atrium.

___ 13. The P wave of the ECG represents the contraction impulse in the atria.

___ 14. Coronary arteries supply the heart with oxygen.

___ 15. In a blood pressure of 130/80, 80 represents the systolic pressure.

___ 16. The medulla of the brain has a region that controls blood pressure.

___ 17. Hypotension is another name for high blood pressure.

___ 18. The spleen is a lymph organ.

___ 19. Leukocytes are more numerous than erythrocytes in human blood.

___ 20. Platelets release clotting factors into the blood.

MULTIPLE CHOICE

21. Blood is best described as a
    a. connective tissue composed of liquid and cells.
    b. muscular pump.
    c. series of passagaes for transport.
    d. system of vessels and nodes.
22. The femoral artery is located in the
    a. abdomen.
    b. arm.
    c. leg.
    d. neck.
23. Select the animal with only two chambers in the heart.
    a. amphibian
    b. bird
    c. fish
    d. reptile
24. Select the incorrect statement about the amphibian heart.
    a. One ventricle is present.
    b. It pumps highly-oxygenated blood to the tissues.

c. Two atria are present.
d. It works with lungs in the adult.

25. In a pulmonary circuit blood flows between the heart and
a. brain.
b. liver.
c. legs.
d. lungs.

26. The main function of the valves in the heart is to
a. develop a low blood pressure.
b. prevent the backflow of blood.
c. produce blood cells in the heart.
d. separate the two sides of the heart.

27. The outermost layer of an artery consists of
a. connective tissue with elastic fibers.
b. smooth muscle tissue.
c. striated muscle tissue
d. thin-walled cells.

28. How many layers of endothelial cells comprise the wall of a capillary?
a. one
b. two
c. three
d. four

29. Select the difference between veins and arteries.
a. The layer of smooth muscle is thicker in veins.
b. Veins have two layers in their walls.
c. Veins have valves.
d. Veins transport blood toward the capillaries.

30. If the heart beats 120 times each minute, the length of one cardiac cycle is about ____________ second.
a. 0.1
b. 0.5
c. 0.8
d. 1.0

31. Another name for systole is
a. closure.
b. contraction.
c. opening.
d. relaxation.

32. The "lub-dub" sounds during the cardiac cycle is due specifically to the
a. closure of valves.
b. leakage of blood.

c. production of blood cells.
d. relaxation of cardiac muscle.

33. The function of the sinoatrial node is to
a. change the size of blood vessels.
b. contract and produce a flow of blood.
c. prevent the backflow of blood.
d. send an impulse to the atria.

34. The T wave of the ECG represents the electrical changes when the
a. atria contract.
b. atria relax.
c. ventricles contract.
d. ventricles relax.

35. Angina pectoralis develops from the
a. blockage of coronary arteries.
b. destruction of lungs tissues.
c. dilation of arteries and veins.
d. production of ant-inflammatory factors.

36. In a blood pressure of 110/70, the figure of 110 represents
a. atrial diastole.
b. atrial systole.
c. ventricular diastole.
d. ventricular systole.

37. Systemic blood pressure is normally lowest in the
a. arteries.
b. arterioles.
c. capillaries.
d. veins.

38. Lymph vessels closely resemble ___________ in structure.
a. arteries
b. arterioles
c. veins
d. yellow marrow

39. ___________ are produced from megakaryocytes.
a. erythrocytes
b. lymphocytes
c. monocytes
d. platelets

40. Hemophilia is a genetic disorder in which the blood
a. decreases its production of plasma.
b. does not clot properly.
c. flows too rapildy.
d. increases its production of erythrocytes.

COMPLETION

41. Fish, amphibians, and reptiles have ___________ hearts, as they are contractile enlargements of vessels.
42. A ___________ bed is a network of microscopic vessels.
43. The ___________ is the large artery transporting blood away from the left ventricle.
44. The Purkinje fibers sends impulses to the ___________ of the heart.

MATCHING

Match each description, by letter, to the correct term. Each choice is used once.

___	45. eosinophil	a. liquid part of the blood
___	46. erythrocyte	b. produces antibodies
___	47. lymphocyte	c. has a major role in blood clotting
___	48. neutrophil	d. biconcave disk
___	49. platelet	e. most abundant kind of white blood cell
___	50. plasma	f. has large, pink-staining granules

ANSWERS

Key Terms

1. A closed circulatory system has a heart pumping blood through an enclosed series of vessels.
2. The ventricle is the larger, more powerful chamber of the heart.
3. The pulmonary circuit sends blood between the heart and lungs.
4. The semilunar valve closes to prevent the backflow of blood from the aorta to the left ventricle.
5. The artery is a vessel transporting blood away from the heart.
6. The capillary is a microscopic vessel that exchanges substances with body cells.
7. The vein is a vessel transporting blood to the heart.
8. The cardiac cycle is a recurring heartbeat.
9. The sinoatrial node is a mass of tissue in the right atrium that signals the atria to contract.
10. The coronary artery sends blood to the heart to supply its needs.
11. Hypertension is high blood pressure.

12. An erythrocyte is a red blood cell that carries oxygen.
13. The leukocyte is a white blood cell that is part of the body's defense system.
14. The platelet is a cell fragment that contributes to blood clotting.
15. Hemophilia is a genetic disorder in which the blood does not clot.

Chapter Concepts

1. Internal transport allows substances to reach cells deep in the organism. This is essential for the multicellular animal where most cells cannot be served from the external environment directly.
2. Homeostasis is the maintenance of the internal environment serving cells. The blood plasma, and interstitial fluid produced from it, are circulatory components. The circulatory system can send more or less blood to an area to carry out homeostasis to serve cells.

Practice Test

1. True	14. True	27. a	40. b
2. True	15. False	28. a	41. lymph
3. False	16. True	29. c	42. capillary
4. True	17. False	30. b	43. aorta
5. False	18. True	31. b	44. ventricles
6. True	19. False	32. a	45. f
7. False	20. True	33. d	46. d
8. False	21. a	34. d	47. b
9. True	22. c	35. a	48. e
10. True	23. c	36. d	49. c
11. False	24. b	37. d	50. a
12. False	25. d	38. c	
13. True	26. b	39. d	

Chapter 22

THE IMMUNE SYSTEM

CHAPTER OVERVIEW

This chapter presents the variety of defense lines for protecting the body against invading microorganisms and their harmful agents. Some immune responses are nonspecific. Specific responses involve the activity of B and T lymphocytes.

The ABO and Rh human blood groups are well-known antigen-antibody systems. Knowledge of these systems prevents clumping of red blood cells. This could occur, for example, in an incompatible blood transfusion.

Disorders of the immune system explain diseases such as autoimmunity and AIDS.

LEARNING OBJECTIVES

After studying this chapter, you should be able to

1. Describe several nonspecific means of defense.
2. List the major steps of the inflammatory reaction.
3. Explain the role of complement in immunity.
4. List, in the correct order, the cascade series of events during antigen-antibody reactions.
5. Describe the makeup of antibody molecules.
6. Explain the basis of memory for immunity.
7. Outline the composition of the ABO blood groups.
8. Outline the composition of the Rh blood groups.
9. Discuss the strategy for successful organ transplantation.
10. Name some disorders of the immune system.

CHAPTER OUTLINE

The Immune System Defends the Body Against Infection

1. Vertebrates have evolved nonspecific defenses.
2. Specific defenses are antibody-mediated and cell-mediated immunity.

The First Responses to Infection Are Nonspecific

1. The skin has several nonspecific responses.
2. The mucus membrane and cilia of the respiratory tract are nonspecific defenses.

Nonspecific Responses Are Activated by the Inflammatory Reaction

1. Damaged cells release histamine which increases capillary blood flow.
2. White cells defeat invading microorganisms.
3. Plasma flows to the damaged area.
4. Monocytes clean up the area.
5. The inflammatory area is a chain of events.

The Complement System Kills Microorganisms Directly

1. The system is a number of proteins that form an MAC.
2. The MAC kills infecting microorganisms.
3. The complement system supplements the inflammatory response.

The Immune Response Is a Specific Defense Against Infection

1. Immunological memory can make it rapid.

An Overview of the Immune Response

1. Antibodies can bind to antigens.
2. T and B Lymphocytes from stem cells mediate this reaction.
3. B cells regulate antibody-mediated immunity.
4. T cells regulate cell-mediated immunity.

Antibody-Mediated Immunity Uses Molecular Weapons

1. Macrophages detect and engulf invading antigens.
2. Marked macrophages encounter helper T cells.
3. T cells activate B cells and form plasma cells and B memory cells.
4. Plasma cells release antibodies.

Antibodies Are Molecular Weapons against Antigens

1. Antibodies are five classes of Y-shaped immunoglobulins.
2. Antibodies recognize and destroy antigens.

T Cells Mediate the Cellular Immune Response

1. Suppressor T cells inhibit the response of B cells and other T cells.
2. Cytotoxic T cells kill cells infected with microorganisms.

The Immune System Has a Memory Function

1. There is resistance from secondary immunity. A second antigenic exposure yields large-scale antibody and T cell production.
2. This is the basis of vaccination.
3. Vaccines are killed or weakened pathogens that stimulate the immune system and establish a memory cell line for future response if antigenic exposure occurs.

Blood Types Are Determined by Cell Surface Antigens

1. The two significant ones are ABO and Rh.

ABO Blood Typing Allows Safe Blood Transfusions

1. There are three alleles in the human population for the ABO blood groups.
2. The A and B alleles each make an antigen with that letter name. The antigens are on the surface of erythrocytes.
3. The antigenic content names the blood type: A, B, AB, or O (no antigens).
4. The antibody content of a person's own blood is the opposite by letter.
5. Clumping of cells occurs in a recipient if the antibodies match transfused antigens of the same letter.

Rh Blood Types Can Cause Immune Reactions between Mother and Fetus

1. The Rh positive type has the Rh antigen. The Rh negative type does not.
2. In HDN an Rh negative mother makes Rh antibodies against the Rh antigen of an Rh positive birth.
3. The mother has made the antibodies by exposure to a former Rh birth.
4. Destruction of these antibodies in the mother before the birth can prevent this.

Organ Transplants Must Be Immunologically Matched

1. An HLA complex is on chromosome number 6.

Successful Transplants Depend on HLA Matching

1. There are a large number of HLA alleles.
2. Close matches improve the chance for an organ transplant survival.

Disorders of the Immune System

1. There is wide range.

Overreaction in the Immune System Causes Allergies

1. They arise from an immune hypersensitivity.
2. IgE antibodies are made against allergens.
3. Secondary immune responses lead to IgE binding to mast cells. This starts an inflammatory response.
4. It can be severe as in anaphylaxis.

Autoimmune Reactions Cause the Immune System to Attack the Body

1. The person's immune system normally does not react against one's own cells.
2. It does happen, however, in an autoimmune response.
3. Addison disease is one example of many known autoimmune diseases.

Immunodeficiency Diseases Impair the Immune System

1. SCID is a rare genetic disorder. It can perhaps be treated by gene therapy.

AIDS Attacks the Immune System

1. The virus kills T4 helper cells.
2. Viral particles are made from these infected cells.

STUDY HINTS

1. Continue previewing a chapter by reading over the key terms and questions/problems before studying the chapter.
2. Try to diagram the steps of the antigen-antibody reactions described in the chapter. Do this in your own words and style.

CHAPTER REVIEW

Key Terms

Define these key terms from the chapter:

1. histamine

2. monocyte

3. antibody

4. antigen

5. macrophage

6. plasma cell

7. T cell

8. secondary immunity

9. haplotype

10. allergen

Chapter Concepts

1. Is there an evolutionary advantage to having a nonspecific defense system in addition to the specific means?
2. There are many examples of how the concepts in this chapter have improved the health of human society. Can you explain some?

Practice Test

TRUE/FALSE

___ 1. The skin is a passive barrier to infectious agents.

___ 2. The release of histamine by damaged cells in an area reduces the capillary blood flow to that area.

___ 3. A monocyte is a type of white blood cell.

___ 4. The complement system works against the inflammatory response.

___ 5. An antigen can cause antibody production.

___ 6. B cells are genetically programmed to produce antigens.

___ 7. Macrophages can engulf invading microorganisms.

___ 8. A plasma cell can synthesize and release many different kinds of antibody molecules.

___ 9. IgD is a type of antigen molecule.

___ 10. Antibody molecules are made up of two different kinds of proteins.

___ 11. Suppressor T cells decrease the immune response of B cells.

___ 12. A vaccine stimulates the production of memory T cells.

___ 13. Each ABO blood type is named by the antibody present.

___ 14. Blood type A contains the B antibody.

___ 15. Blood type O contains the A and B antibodies.

___ 16. HDN can develop when an Rh positive mother has an Rh negative fetus.

___ 17. The HLA alleles are located on human chromosome number 6.

___ 18. Allergens are weak antigens.

___ 19. Myasthenia gravis is not an autoimmune disease.

___ 20. AIDS is caused by a disease-producing bacterium.

MULTIPLE CHOICE

21. Which of the following actions is not part of the inflammatory response?
    a. Blood flow increases to a damaged area.
    b. Damaged cells release histamine.
    c. Macrophages migrate to a damaged area.
    d. Plasma exits from the injured cells.
22. Complement consists of a number of different
    a. carbohydrates.
    b. lipids.
    c. nucleic acids.
    d. proteins.
23. Opsonization of bacteria means that they are
    a. coated with a substance.
    b. removed from an area.
    c. stimulated to reproduce.
    d. transported in the blood.
24. Antibodies are mainly ___________ molecules.
    a. carbohydrate
    b. lipid
    c. nucleic
    d. protein
25. T cells are originally made in the
    a. bone marrow.
    b. liver.
    c. spleen.
    d. thymus gland.
26. Humoral immunity is
    a. antibody-mediated.
    b. antigen-mediated.

c. platelet-mediated.
d. thymus-inhibited.

27. A helper T cell is a kind of
a. eosinophil.
b. lymphocyte.
c. monocyte.
d. neutrophil.

28. Immunoglobulins are
a. antibodies.
b. antigens.
c. lymphocytes.
d. neutrophils.

29. The immunoglobulins are ___________ -shaped.
a. A
b. B
c. Y
d. Z

30. Cytotoxic cells are ___________ T cells.
a. antigenic
b. helper
c. killer
d. suppressor

31. ___________ T cells remain in the body awaiting the reintroduction of an antigen.
a. cytotoxic
b. helper
c. memory
d. suppressor

32. A vaccine contains a
a. disease-causing antibody.
b. disease-causing antigen.
c. potent enzyme.
d. potent lipid.

33. During inflammation, which of the following steps do not occur?
a. Break down of a wall around fibrin clot.
b. Arrival of neutrophils.
c. Engulfment of invading bacteria.
d. Secretion of chemical mediators.

34. There are ___________ alleles in the human population for the ABO blood groups.
a. two
b. three

c. six
d. eight

35. Type O blood can safely receive blood type
    a. A.
    b. B.
    c. AB.
    d. O.
36. The antibody content in the blood of a person with type O blood is
    a. A only.
    b. B only.
    c. A and B.
    d. neither A nor B.
37. HDN develops with an
    a. Rh+ mother and Rh+ fetus.
    b. Rh+ mother and Rh– fetus.
    c. Rh– mother and Rh+ fetus.
    d. Rh– mother and Rh– fetus.
38. ___________ antibodies are most active in allergic reactions.
    a. IgA
    b. IgE
    c. IgM
    d. IgX
39. Which of the following diseases is not an autoimmune disease?
    a. Addison disease
    b. Grave's disease
    c. hemophilia
    d. polymyositis
40. During HIV infection, there is a gradual decrease in ___________ helper cells.
    a. T2
    b. T4
    c. T6
    d. T8

COMPLETION

41. The respiratory tract is lined with ___________ membranes.
42. In an inflamed area of the body, the temperature changes by ___________ .
43. Lymphocytes are produced from the mitosis of ___________ cells in the bone marrow.
44. ___________ cells control cell-mediated immunity.

45. An antibody consists of H and __________ chains.
46. The __________ immune response is the basis for a vaccination, as memory cells become active if a person is re-exposed to a disease-causing agent.
47. Type B blood contains the __________ antibody.
48. The success of skin grafts depends on matching __________ antigens.
49. IgE antibodies bind to __________ cells during an allergic reaction.
50. Scleroderma is an __________ disease.

ANSWERS

Key Terms

1. Histamine is a substance that is produced by damaged blood vessels. It can dilate blood vessels.
2. A monocyte is a white blood cell which cleans up dead cells and microorganisms.
3. An antibody is a molecule made by an organism that can bind to a foreign molecule and destroy it.
4. An antigen is a foreign molecule that invades the body and could harm it unless destroyed.
5. A macrophage is a large, wandering white blood cell that can engulf antigens and foreign cells.
6. A plasma cell is formed from a B lymphocyte. It can make many different antibodies.
7. A T cell is the cell involved in cell-mediated immunity. It has many types.
8. A secondary immune response occurs with a re-exposure to an antigen. It is made from a memory cell line.
9. A haplotype is an array of alleles on human chromosome number 6.
10. An allergen is a weak antigen that can stimulate an allergic reaction.

Chapter Concepts

1. Nonspecific reactions can ward off many invading agents, preventing them from entering and infecting the body. Therefore, specific responses are not necessary. This is a saving of resources and energy for the body.
2. Examples include the prevention of HDN, vaccinations to prevent disease, and successful organ transplants.

Practice Test

1. True
2. False
3. True
4. False
5. True
6. False
7. True
8. True
9. False
10. True
11. True
12. True
13. False
14. True
15. True
16. False
17. True
18. True
19. False
20. False
21. d
22. d
23. a
24. d
25. a
26. a
27. b
28. a
29. c
30. c
31. c
32. b
33. a
34. b
35. d
36. c
37. c
38. b
39. c
40. b
41. mucus
42. increasing
43. stem
44. T
45. L
46. secondary
47. A
48. histocompatibility
49. mast
50. autoimmune

☙ Chapter 23 ❧

DIGESTION AND NUTRITION

CHAPTER OVERVIEW

This chapter provides extensive information on the digestive system. Many food molecules, such as the polysaccharide starch, cannot be used in their original form when eaten. They must be chemically changed into smaller subunits for use in body metabolism. The digestive system functions to convert food molecules into a chemical form that makes them usable.

There is relationship between digestion and nutrition. The products of digestion provide the nutrients that enter cells for roles ranging from energy storage to synthesis of important cellular substances.

LEARNING OBJECTIVES

After studying this chapter, you should be able to

1. Compare several kinds of feeding mechanisms.
2. Outline the steps of food processing.
3. Describe the structure and functions of each component of the human digestive system.
4. List the products of the chemical digestion of carbohydrates, fats, and proteins.
5. Explain how appetite is regulated.
6. Explain how digestion is hormonally regulated.
7. State the relationship between digestion and nutrition.
8. Outline the makeup of a well-balanced diet.

CHAPTER OUTLINE

Feeding and Digestion Process Food into Nutrients

1. These are necessary specializations in multicellular animals.

Animals Use a Variety of Feeding Mechanisms

1. Many animals are ingestive eaters.
2. Parasites are absorptive feeders.
3. Other animals can be filter, substrate, and fluid feeders.

Animal Digestive Systems Have Two Major Features

1. One is a sac-like or tube-within-a-tube body plan.
2. The other is intracellular or extracellular digestion.

Food Processing and Digestion Occur in Stages

1. The stages are movement, secretion, digestion, absorption, and elimination.
2. They occur as food passes through a series of compartments.

Components of the Human Digestive System

1. Food passes through a series of compartments of a tract. The order is: mouth—pharynx—esophagus—stomach—small intestine—large intestine.
2. Accessory organs are the salivary glands, pancreas, and the biliary system.

The Mouth and Pharynx: Feeding and Swallowing

1. In the mouth of vertebrates there is the tearing, chewing, and grinding of food.
2. The tongue contains chemical receptors for taste.

Digestion Begins in the Mouth

1. Salivary amylase catalyzes the conversion of starch into sugar.
2. Mucus moistens the food.
3. Bicarbonate ions are present in the saliva

The Digestive System and Respiratory System Open into the Pharynx

1. The tongue moves a bolus into the pharynx.
2. This triggers an involuntary swallowing reflex.
3. The peristalsis of the esophagus moves food into the stomach.
4. A sphincter muscle prevents food from moving back into the stomach.

The Stomach Stores Food and Breaks It Down

1. The stomach has a tremendous capacity for food storage.

The Stomach Stores, Mixes, and Digests Food

1. Gastric pits secrete juice.
2. Gastric secretions are controlled.
3. Under great acidity pepsinogen is converted into pepsin to start the chemical digestion of proteins.

Two Digestive Processes Take Place in the Stomach

1. An acid chyme leaves the stomach by peristalsis.
2. Alcohol and a few other compounds are absorbed.

Gastric Juices Can Digest the Stomach Lining

1. Secreted mucus protects this epithelial level.
2. Bicarbonate ions buffer acidity.
3. The breakdown of protective mechanisms can lead to an ulcer.

The Small Intestine: The Major Site of Digestion and Absorption

1. The pyloric sphincter prevents the backflow of food into the stomach.

Chemical Digestion Is Completed in the Small Intestine

1. The chyme first enters the duodenum.
2. Epithelial cells secrete mucus.
3. The pancreas secretes enzymes and an alkaline solution.
4. Bile enters from the gallbladder through the bile duct. It is necessary to emulsify fats.
5. The chemical digestion of carbohydrates and proteins is continued and finished here. Fats are started and finished.

The Small Intestine Is Well Adapted to Absorb Nutrients

1. Most occurs in the jejunum.
2. By absorption, the molecules of digestion, such as monosaccharides and amino acids, cross the walls of villi and enter the blood.
3. In lactose intolerance, the necessary enzyme is not present.
4. Micelles help fat absorption, which is completed in the ileum.

The Large Intestine Absorbs Water and Stores Waste

1. It consists of the colon, cecum, appendix, and rectum.
2. There are small involuntary movements and alkaline secretions.
3. Some absorption occurs.
4. The feces is formed and eliminated.

Regulation of Appetite, Digestion, and Nutrients

Appetite Is Controlled by Internal and External Factors

1. The hypothalamus has centers that regulate appetite.

Digestion Is Hormonally Regulated

1. Gastrin stimulates the release of pepsinogen and HCl.
2. Secretin stimulates pancreatic secretions.
3. CCK signals the gallbladder to release bile.

Storage and Release of Glucose for Metabolism Are Hormonally Controlled

1. Extra glucose in the blood is converted to glycogen for storage.
2. Several hormones interact to control glucose in the blood.

Nutrition and the Human Diet

1. Organisms can be chemotrophs, phototrophs, and heterotrophs.
2. Animals are heterotrophs.

The Well-Balanced Diet

1. They include the carbohydrates, lipids, and proteins.

Carbohydrates Should Be the Primary Energy Source

1. They are in a variety of foods.

Proteins Provide Essential Amino Acids

1. They have many functions in metabolism.
2. Some amino acids are essential.

Lipids Serve as a Method for Energy Storage

1. They have the highest energy yield and other functions, such as insulation.

A Well-Balanced Diet Includes Vitamins And Minerals

Vitamins Are Required to Carry Out Metabolic Reactions

1. Vitamins act as coenzymes and cofactors.
2. Vitamins are water-soluble and fat-soluble.

Minerals Are Required for Many Functions

1. Functions range from hormone to hemoglobin production.

Diet and Its Links to Disease

1. Diseases include cardiovascular diseases.
2. Studies are underway to learn more about these links.

STUDY HINTS

1. Review the information of chemistry that you learned in Chapter Two. It is applicable to the information in this chapter?

2. Review previous chapters on body structure and function. Can you now understand how several organ systems work together. One example is the relationship between the circulatory, respiratory, and digestive systems.

CHAPTER REVIEW

Key Terms

Define these key terms from the chapter:

1. ingestive
2. salivary amylase
3. bolus
4. peristalsis
5. acid chyme
6. duodenum
7. jejunum
8. villi
9. hypothalamus
10. gastrin
11. heterotroph

12. essential amino acid

Chapter Concepts

1. The different branches of science relate. One branch can be a useful tool for understanding another. How do the principles of chemistry relate to an understanding of this chapter?
2. Continuing the theme of homeostasis, what do you think is one example cited in this chapter?

Practice Test

TRUE/FALSE

___ 1. Ingestive eaters feed only on plants.

___ 2. Sponges carry out intracellular digestion.

___ 3. The duodenum is part of the large intestine.

___ 4. Starches are broken down chemically through the action of salivary amylase.

___ 5. Peristalsis is the passage of digested molecules from the digestive tract into the blood.

___ 6. Gastric secretions are controlled by the nervous system and endocrine system.

___ 7. An acid environment is necessary for the conversion of pepsinogen into pepsin.

___ 8. A large percentage of peptic ulcers are caused through the infection by a bacterium.

___ 9. The pancreas produces bile.

___ 10. The pancreas produces enzymes that are secreted into the small intestine.

___ 11. The presence of villi decreases the surface area of the mucous lining of the small intestine.

___ 12. Starch is chemically changed into maltose by the action of pancreatic amylase.

___ 13. Through a lactose intolerance too much lactose is broken into two monosaccharides.

___ 14. Gluten enteropathy affects the absorption of nutrients.

___ 15. Peptide fragments are the final products of digestion of proteins.

___ 16. Some chemical breakdown of carbohydrates occurs in the large intestine.

___ 17. Both the appetite and satiety centers are located in the hypothalamus.

___ 18. CCK is a hormone released into the blood in response to the arrival of fats in the duodenum.

___ 19. Most chemotrophs are bacteria.

___ 20. Tryptophan is not an essential amino acid in humans.

MULTIPLE CHOICE

21. Oysters are ____________ feeders.
    a. autotroph
    b. filter
    c. fluid
    d. substrate
22. After the mouth, food in humans passes through the three digestive compartments in the correct order of
    a. esophagus—pharynx—stomach.
    b. pharynx—esophagus—stomach.
    c. large intestine—stomach—small intestine.
    d. stomach—esophagus—large intestine.
23. Salivary amylase works in the
    a. large intestine.
    b. mouth.
    c. small intestine.
    d. stomach.
24. A bolus is a(n)
    a. mass of food.
    b. muscle contracting in the small intestine.
    c. nerve signaling the stomach.
    d. signal acting as a hormone.
25. Peristalsis depends on the
    a. absence of enzymes.
    b. absence of hormones.
    c. contraction of cardiac cells.
    d. contraction of smooth muscle fibers.

26. The colon is part of the
    a. large intestine.
    b. mouth.
    c. small intestine.
    d. stomach.
27. The activity of pepsin begins the chemical breakdown of
    a. carbohydrates.
    b. lipids.
    c. nucleic acids.
    d. proteins.
28. Acid chyme is produced in the
    a. large intestine.
    b. mouth.
    c. small intestine.
    d. stomach.
29. Most absorption of nutrients occurs in the
    a. esophagus.
    b. large intestine.
    c. small intestine.
    d. stomach.
30. The pyloric sphincter regulates the movement of food into the
    a. esophagus.
    b. pharynx.
    c. small intestine.
    d. stomach.
31. Normally, food passes through the small intestine in the order
    a. duodenum—ileum—jejunum.
    b. duodenum—jejunum—ileum.
    c. ileum—duodenum—jejunum.
    d. ileum—jejunum—deudenum.
32. The function of bicarbonate ions in the small intestine is to
    a. make sugars.
    b. neutralize acidity.
    c. serve as a hormone.
    d. store enzymes.
33. The stomach is lined internally with a ____________ membrane.
    a. cutaneous
    b. mucus
    c. serous
    d. plasma

34. Starch is chemically broken into the disaccharide
    a. fructose.
    b. lactose
    c. maltose
    d. sucrose
35. Bile is necessary to promote the digestion of
    a. carbohydrates.
    b. fats.
    c. glycoproteins.
    d. proteins.
36. The main function of the colon is
    a. absorption.
    b. chemical digestion.
    c. elimination of wastes.
    d. storage of nutrients.
37. Gastrin signals the
    a. absorption of nutrients.
    b. elimination of wastes.
    c. increase in the hunger sensation.
    d. release of HCl and pepsinogen.
38. Secretin is a(n)
    a. amino acid.
    b. enzyme.
    c. hormone.
    d. monosaccharide.
39. The ___________ is one organ that stores extra glucose from the blood.
    a. brain
    b. liver
    c. pancreas
    d. small intestine
40. The metabolism of ___________ generates the highest energy yields.
    a. disaccharides
    b. fats
    c. polysaccharides
    d. proteins

COMPLETION

41. Heartburn develops from the entrance of ___________ juice into the esophagus.

42. Microvilli establish a ___________ border along the inner lining of the small intestine.

43. Bile salts surround fat molecules to form ___________ which are taken into epithelial cells.

44. Vitamins are components of ___________ and cofactors in cells.

MATCHING

Match each description, by letter, to the correct vitamin or mineral. Each choice is used once.

___ 45. A	a. fat-soluble vitamin needed to make visual pigment
___ 46. B-1	b. deficiency of this vitamin causes pellagra
___ 47. Ca	c. vitamin needed for blood clotting
___ 48. Fe	d. mineral important for bone and tooth formation
___ 49. K	e. mineral which is a part of hemoglobin
___ 50. niacin	f. deficiency of this vitamin causes beriberi

ANSWERS

Key Terms

1. Ingestive means to eat plants or animals or both.
2. Salivary amylase is an enzyme that catalyzes the chemical conversion of starch into the sugar maltose.
3. A bolus is the swallowed food ball.
4. Peristalsis is a wave of muscle contraction that moves food through the digestive tract.
5. Acid chyme is a mixture of food entering the duodenum from the pyloric end of the stomach.
6. The duodenum is the first segment of the small intestine receiving food.
7. The jejunum is between the duodenum and ileum segments of the small intestine.
8. Villi are finger-like projections of the small intestine inner lining.
9. The hypothalamus is a brain region with many functions, including the control of appetite.
10. Gastrin is a hormone that signals the release of HCl and pepsinogen into the stomach.
11. A heterotroph eats plants and animals for energy.
12. An essential amino acid must be supplied in the diet, as the organism cannot make it.

Chapter Concepts

1. The large molecules in the diet undergo chemical digestion. Macromolecules are broken into subunits. Their chemical bonds are broken. Therefore, chemical bonds and families of biologically important molecules are very relevant to the action occurring during digestion.
2. One example is the control of nutrients in the blood. Glucose must be regulated. Digestion supplies the raw materials for that regulation.

Practice Test

1. False
2. True
3. False
4. True
5. False
6. True
7. True
8. True
9. False
10. True
11. False
12. True
13. False
14. True
15. False
16. False
17. True
18. True
19. True
20. False
21. b
22. b
23. b
24. a
25. d
26. a
27. d
28. d
29. c
30. c
31. b
32. b
33. b
34. c
35. b
36. c
37. d
38. c
39. b
40. b
41. gastric
42. brush
43. micelles
44. coenzymes
45. a
46. f
47. d
48. e
49. c
50. b

☙ Chapter 24 ❧

EXCRETION AND TRANSPORT

CHAPTER OVERVIEW

This chapter sets forth more examples of homeostasis. Metabolic products must be eliminated from the organism. The composition of the extracellular fluid (internal environment) of the organism must be regulated. This cannot occur just on the input side from ingested nutrients, for example. A balance must also include the output side from the response of excretory structures.

The control of the internal environment is refined by the vertebrate kidney. The kidneys receive a large amount of blood each minute. Through the processes of filtration, tubular reabsorption, and tubular secretion the blood plasma and interstitial fluid (both parts of the extracellular fluid) are regulated for a large array of substances. These processes are conducted by functional units in the kidneys, the nephrons. The kidneys also have other functions, such as regulating blood pressure.

LEARNING OBJECTIVES

After studying this chapter, you should be able to

1. Compare the several ways animals have evolved means to eliminate nitrogen.
2. Contrast the processes of osmoregulators and osmoconformers.
3. State the problems of water and mineral balance faced by terrestrial, marine, and freshwater organisms.
4. Compare the excretory structures of several invertebrates.
5. Describe the location and function of the human excretory system.
6. Describe the components of the nephron.
7. Explain the processes of filtration, tubular reabsorption, and tubular secretion.
8. State the functions of several hormones regulating water and salt balance.

9. List some functions of the kidney in addition to controlling the composition of the extracellular fluid.
10. Discuss how kidney functions can be disrupted.

CHAPTER OUTLINE

Excretory Systems Eliminate Excess Water, Salts, and Wastes

1. These substances are the by-products of metabolism.
2. The elimination is essential for the survival of multicellular organisms.

Excretion Regulates the Composition of Extracellular Fluids

1. The components of the vertebrate system are: kidneys, liver, lungs, and skin.

Excretion and Homeostasis

1. Excretion involves the movement of waste products across plasma membranes.
2. Elimination is the removal of undigested food.

Nitrogen Wastes Are Produced in Protein Metabolism

1. An amino group can be removed and used to form ammonia.

Animals Use Several Methods of Nitrogen Excretion

1. Ammonia usually is eliminated into water.
2. Urea is made from ammonia and eliminated by many land animals. It is made by the liver and eliminated by the kidneys. It requires little water for elimination.
3. Uric acid can be eliminated with little water loss. It is excreted by reptiles, birds, and adult insects.
4. Uric acid is also produced and stored in the egg of birds and reptiles.

Water and Salt Balance Is an Important Function of Excretion

1. Aquatic animals conduct osmoregulation.
2. Osmoconformers reach a balance with their isotonic external environment.
3. Osmoregulators control water and dissolved salts in their bodies. This is independent of the external environment.
4. Marine fish face the problem of losing water to their external environment and must osmoregulate.
5. Freshwater fish face the opposite problem encountered by marine fish. They must also osmoregulate.

Excretory Systems Perform Three Main Functions

1. They collect and filter body fluids.
2. They remove and concentrate body fluid wastes.

3. They eliminate excretory products.

Invertebrates Have Tubular Excretory Organs
1. Flatworms have a nephridium with flame cells.
2. Insects have Malpighian tubules.

Vertebrates Have Paired Kidneys for Excretion
1. They are one type of tubular organ.

The Human Excretory System Has Four Main Components

Overview of the System
1. The functional unit is the nephron.
2. Urine passes from each kidney through a ureter.
3. The ureters send urine to the bladder.
4. The urethra drains the bladder.

The Nephron and How It Works
1. It consists of a glomerulus and renal tubule.
2. Blood entering the kidney passes through the glomerulus of each nephron.
3. From there fluid passes through the tubular part of the nephron in the following order: proximal tubule—loop—distal tubule.
4. Water and solutes leave the blood from the glomerulus by filtration.
5. Through tubular reabsorption, water and substances return to the blood in the peritubular capillaries.
6. Through tubular secretion, substances leave the peritubular capillaries and enter the nephron tubule near the end.
7. Filtration, tubular reabsorption, and tubular secretion are processes regulating the concentration of substances in the blood such as water and sodium.

Kidney Function Regulates Internal Fluid Composition
1. They regulate the extracellular fluid and blood pressure.
2. They maintain ionic and acid-base balance.
3. They excrete metabolic by-products.

Water and Salt Levels Are Controlled by Hormones
1. ADH from the pituitary signals the nephron tubules to reabsorb water.
2. Aldosterone signals the nephron tubules to reabsorb sodium.
3. Angiotensin II constricts blood vessels and controls blood pressure. Renin from the kidney forms this hormone.

Kidney Function Can Be Disrupted by a Number of Factors
1. Some symptoms can be treated by dialysis.
2. Kidney transplantation is another option.

STUDY HINTS

1. Review the kinds of cell transport from Chapter Three. How many are relevant to your understanding of the functions discussed in this chapter?
2. Can you write your own summary of this chapter? Use your own words rather than simply repeating the summary at the end of the textbook chapter.

CHAPTER REVIEW

Key Terms

Define these key terms from the chapter:

1. osmoregulation
2. osmoconformer
3. nephridium
4. Malpighian tubule
5. nephron
6. ureter
7. urethra
8. glomerulus
9. filtration
10. tubular reabsorption

11. tubular secretion

12. ADH

13. aldosterone

14. renin

15. angiotensin II

Chapter Concepts

1. How do the study of chemistry, cell transport, and circulation relate to an understanding of kidney function?
2. Can you select another concept that you have studied (and will study) that relates to the physiology (study of function) in this chapter?

Practice Test

TRUE/FALSE

___ 1. Excretion is the same process as elimination.

___ 2. Nitrogen wastes are a product of the metabolism of proteins.

___ 3. Urea is the major nitrogen metabolic product of adult amphibians.

___ 4. Uric acid is not a major nitrogen metabolic product of reptiles.

___ 5. Osmoconformers do not have a system of osmoregulation.

___ 6. Freshwater vertebrates face the problem of gaining water from their external environment.

___ 7. Animals that live on land face the problem of losing water to their external environment.

___ 8. The nephridium is the excretory organ in insects.

___ 9. The urethra transports urine from the kidney to the bladder.

___ 10. The glomerulus is a capsule that surrounds a network of capillaries in the nephron.

___ 11. The renal artery transports blood to the kidney.

___ 12. Because of filtration, substances do not exit from the blood in the glomerulus.

___ 13. Substances return to the blood through reabsorption.

___ 14. Tubular secretion occurs mainly at the proximal convoluted tubule.

___ 15. Substances enter the blood by tubular secretion.

___ 16. The loop of Henle participates in a countercurrent exchange.

___ 17. ADH is a hormone produced by the kidney.

___ 18. An increase in the secretion of ADH increases the permeability of the nephron tubules to water.

___ 19. Aldosterone is secreted by the adrenal glands.

___ 20. An increase in the secretion of angiotensin II dilates blood vessels.

MULTIPLE CHOICE

21. Which of the following choices is an excretory organ?
    a. hypothalamus
    b. kidney
    c. liver
    d. skin
22. Amino groups are removed from subunits in ___________ metabolism.
    a. fat
    b. monosaccharide
    c. polysaccharide
    d. protein
23. Urea is filtered at the glomerulus by the
    a. kidney.
    b. large intestine.
    c. liver.
    d. small intestine.
24. Which of the following life forms excretes substantial amounts of uric acid?
    a. adult insects
    b. birds
    c. mammals
    d. reptiles

25. Osmoconformers live in __________ external environments.
    a. hypertonic
    b. hypotonic
    c. isotonic
    d. terrestrial
26. Animals that live in a hypertonic external environment face the problem of water
    a. blockage.
    b. destruction.
    c. gain.
    d. loss.
27. The __________ sac in the reptile egg fills with metabolic wastes.
    a. allantois
    b. amnion
    c. chorion
    d. yolk
28. The flame cell is found in the
    a. kidney.
    b. liver.
    c. nephridium.
    d. urethra.
29. Malpighian tubules are excretory organs of
    a. birds.
    b. insects.
    c. mammals.
    d. reptiles.
30. From the kidney urine normally passes from the
    a. bladder to the ureter to the urethra.
    b. bladder to the urethra to the ureter.
    c. ureter to the bladder to the urethra.
    d. urethra to the bladder to the urethra.
31. Between the proximal and distal tubules, fluid passes through the
    a. Bowman's capsule.
    b. glomerulus.
    c. loop of Henle.
    d. urethra.
32. Which substance normally does not pass through the pores of the glomerulus during filtration?
    a. amino acids
    b. antibodies
    c. glucose
    d. sodium ions

33. Which capillaries are involved with reabsorption in the nephron?
    a. cortical
    b. glomerulus
    c. peritubular
    d. renal
34. ____________ ions undergo tubular secretion to regulate the acid-base balance in the blood.
    a. hydrogen
    b. iron
    c. phosphate
    d. sodium
35. If the kidneys filter 192 liters of water over 24 hours and reabsorb 189.5 liters, the amount of urine produced is ____________ liters.
    a. 1.5
    b. 2.5
    c. 3.5
    d. 194.5
36. The ____________ of the brain monitors the osmotic concentration of the blood.
    a. cerebrum
    b. hypothalamus
    c. medulla
    d. pons
37. ADH acts to
    a. decrease metabolic reaction rates.
    b. decrease water reabsorption.
    c. increase metabolic reaction rates.
    d. increase water reabsorption.
38. If the concentration of sodium in the blood decreases ____________ is released by the adrenal cortex.
    a. aldosterone
    b. epinephrine
    c. norepinephrine
    d. sodium
39. Renin acts on a blood protein called
    a. aldosterone.
    b. angiotensinogen.
    c. erythropoietin.
    d. ferrin.
40. Bladder infections are nearly always caused by
    a. bacteria.
    b. fungi.

c. protozoa.
d. viruses.

COMPLETION

41. Cells produce the gas __________ as a waste product of metabolism.
42. __________ pigments are formed by the breakdown of hemoglobin in the liver.
43. __________ fluids directly surround the embryo in the reptile egg.
44. The __________ is the microscopic, functional unit of the kidney.
45. The ureter contains __________ muscle tissue in its wall.
46. __________ is the first step of urine formation and occurs in the glomerulus.
47. Most reabsorption occurs in the __________ tubule of the nephron.
48. __________ is the main nitrogen waste product in humans.
49. When sodium is reabsorbed water follows it back into the blood by the transport process of __________ .
50. __________ is the infection of the nephrons.

ANSWERS

Key Terms

1. Osmoregulation is the maintenance of an internal environment of the organism that is independent of the external environment.
2. An osmoconformer maintains an internal environment that matches the makeup of the external environment.
3. The nephridium is the excretory organ of the flatworm.
4. The Malpighian tubule is the excretory organ of the insect.
5. The nephron is the functional unit of the kidney.
6. The ureter transports urine from the kidney to the bladder.
7. The urethra drains urine from the bladder.
8. The glomerulus is a group of capillaries inside the Bowman's capsule of the nephron.
9. Filtration is the mechanism by which substances leave the blood of the glomerulus.
10. Tubular reabsorption is the mechanism by which substances return to the blood (peritubular capillaries) from the tubular part of the nephron.

11. Tubular secretion is the mechanism by which substances leave the blood of the peritubular capillaries and enter the tubular part of the nephron near the end.
12. ADH is a hormone that signals the nephron tubule to reabsorb water.
13. Aldosterone is a hormone that signals the nephron tubule to reabsorb sodium.
14. Renin is made by the kidney to form angiotensin II.
15. Angiotensin II is a hormone that constricts blood vessels and raises blood pressure. It also stimulates sodium reabsorption.

Chapter Concepts

1. Chemical substances must be regulated. The balance of body chemistry is a function of the excretory system. Various transport processes account for the removal and retention of substances through the body. The circulating blood brings these substances to the kidneys for monitoring and regulation.
2. Another system is the endocrine system and the concept of signaling through hormones. Consider the roles of ADH and aldosterone on kidney functions. You were introduced to the endocrine system in Chapter 18. Look ahead to Chapter 26.

Practice Test

1. False	14. False	27. a	40. a
2. True	15. False	28. c	41. carbon dioxide
3. True	16. True	29. b	42. bile
4. False	17. False	30. c	43. amnion
5. True	18. True	31. c	44. nephron
6. True	19. True	32. b	45. smooth
7. True	20. False	33. c	46. filtration
8. False	21. a	34. a	47. proximal
9. False	22. d	35. b	48. urea
10. False	23. a	36. b	49. osmosis
11. True	24. c	37. d	50. pyelonephritis
12. False	25. c	38. a	
13. True	26. d	39. b	

Chapter 25

NERVOUS SYSTEM AND SENSORY SYSTEM

CHAPTER OVERVIEW

This chapter emphasizes the need for signaling and integration in the animal body. Stimuli in the external environment must be sensed. Responses must be made to these environmental changes. Through the evolving groups of animals the nervous system has developed an increasingly complex form to meet the needs of the organism for survival.

The neuron is the cell that sends a signal in the nervous system. Neurons form circuits in the organism. In vertebrates the nervous system is organized into central and peripheral divisions. The brain of the CNS is most complex in the human. Working with receptors that sense stimuli and effectors that make responses, the nervous system establishes the ability for complex behavior in the organism.

LEARNING OBJECTIVES

After studying this chapter, you should be able to

1. Compare the nervous systems in different animals for complexity.
2. Distinguish among the different kinds of nerve cells.
3. Explain the events of signaling along the neuron.
4. Explain the events of signaling at the synapse.
5. Outline the makeup of the nervous system in the vertebrate.
6. List the structures and accompanying functions of the regions of the brain.
7. Describe the structures and functions of the spinal cord.
8. Compare the different kinds of receptors.
9. Explain the sensory functions of vision, balance, and vision.
10. Describe the structure and functions of the peripheral nervous system.

CHAPTER OUTLINE

The Nervous System: Sensation, Integration, and Coordination

1. For homeostasis, organisms must sense changes in the environment and make responses.

Nervous Systems Perform Three Basic Functions

1. They receive sensory input, integrate this input, and respond to stimuli.
2. There is a variety of sensory input.
3. Integration generates a response.
4. The response is a motor output.
5. The nervous and endocrine systems control the responses. They do this through different mechanisms.

Nervous Systems Have Two Divisions

1. One division is the central nervous system, the brain and spinal cord.
2. The other division is the peripheral nervous system, composed of nerves.
3. Neurons are the functional units of this system.

Nervous Systems Have Several Levels of Organization

1. Large, mobile animals often have complex and well-organized nervous systems.

Nerve Nets Are Simple Nervous Systems

1. They are present is coelenterates, cnidarians, and echinoderms.
2. These organisms have radial symmetry and carry out complex responses.

Bilateral Symmetry Is Associated with More Complex Nervous Systems

1. It is accompanied by cephalization.
2. Ganglia form a small brain in the flatworm.
3. Vertebrates have a highly-developed CNS.

Neurons Are the Functional Units of the Nervous System

1. Neurons are protected by connective tissue.
2. Glial cells protect and support neurons.

Neurons Have Distinctive Structural Features

1. Sensory neurons send signals toward the CNS.
2. Motor neurons send signals away from the CNS and toward effectors.
3. Interneurons complete the signal circuit between sensory and motor neurons.
4. Dendrites and axons are processes of the neuron.
5. The axon carries the signal away from the neuron cell body.

6. Some axons are covered with a myelin sheath from Schwann cells. The nodes of Ranvier are breaks in the myelin.
7. Impulses jumping from node to node travel more rapidly.

Neurons Maintain an Electrical Potential across Their Membranes

1. The neuron is positive on the outside of the plasma membrane during the resting membrane potential.
2. At rest sodium ions are pumped to the outside and potassium ions to the inside.

Nerve Signals Cause a Temporary Change in Resting Potential

1. An action potential is triggered by the inflow of sodium ions through plasma membrane channels.
2. This sodium influx reverses the polarity of the neuron.
3. An outflow of potassium ions milliseconds later reestablishes the resting membrane potential.
4. The cell pumps sodium out of the cell and potassium into the cell to restore their original concentration gradients.

Action Potentials Move along the Neuron

1. The action potential is the electrical reversal from the influx of sodium. This change travels along the plasma membrane of the neuron.
2. The pumping of sodium out of the cell and potassium into the cell restores the original distribution of ions.
3. An action potential travels from the dendrites to the cell body to the axon.
4. The presence of myelin with nodes along the axon increases the speed of impulse transmission.

Signals Are Transferred between Neurons at Synapses

1. The synapse is the junction between neurons.
2. The signal at the synapse is chemical.
3. A neurotransmitter diffuses from the axon of one cell across the synapse and contacts receptors on the next neuron in the circuit.
4. This excites the next neuron, altering the permeability of the cell and producing an action potential.

Chemicals and Diseases Affect the Nervous System at Synapses

1. Parkinson's disease is from a deficiency of dopamine in the brain.
2. *Clostridium tetani* prevents the release of GABA.
3. Drugs often alter synaptic transmission.

An Overview of the Vertebrate Nervous System

1. It consists of the central and peripheral divisions.

2. The motor pathways are somatic and autonomic.
3. The autonomic system has a sympathetic and parasympathetic systems.

Vertebrate Brains Have a High Capacity for Signal Integration

1. The CNS forms from a tube which enlarges into three hollow swellings. The brain and spinal cord form from these swellings.
2. Brain size has increased progressively through the evolving classes of vertebrates.
3. The brain has been subdivided into a hindbrain, midbrain, and forebrain.
4. The forebrain and its cerebrum has increased in size, particularly in humans where it affects behavior.

Structure and Function of the Human Brain: An Overview

1. The regions are the brain stem, cerebellum, and the forebrain with the diencephalon and cerebrum.

The Brain Stem Is the Most Primitive Part of the Brain

1. It consists of the medulla oblongata and pons. Brain stem functions include sensory-motor relay and control of heart rate, blood vessel size, respiration, and many other responses.
2. The midbrain carries out relay and coordinates sensory functions for vision.

The Cerebellum Coordinates Body Movement

1. Functions include fine movement coordination, posture, and balance.

The Forebrain Is a Sophisticated Processing Center

1. The diencephalon is the lower forebrain. Its thalamus is a sensory switching center. Its hypothalamus has roles ranging from integration with the endocrine system to emotion and behavior.

The Cerebrum is the Largest Portion of the Human Brain

1. The cerebrum is the higher forebrain.
2. The corpus callosum connects its two hemispheres.
3. The cortex is the outer gray layer. It overlays white matter.

The Cerebral Cortex Is a Complex Mosaic of Functional Regions

1. The lobes of the cerebral cortex are the occipital, temporal, parietal, and frontal.
2. Functions can be mapped on these lobes.

The Lobes of the Cerebral Cortex Are Specialized for Different Functions

1. The occipital lobe has visual sensory areas.
2. The temporal lobe receives auditory signals.

3. The parietal lobe has many mapped sensory areas for body regions.
4. The frontal lobe controls skeletal muscle activity and integration, speech, and thought processes.
5. One-half of the cortex is association areas.

The Spinal Cord Links the Brain and the Body

1. It has an inner zone of gray matter and outer white matter.
2. Spinal nerves connect to it at intervals.
3. The cord mediates the reflex arc and is a vertical link between the brain and body regions.

The Peripheral Nervous System: An Overview

1. It has sensory (afferent) and motor (efferent) pathways

The Peripheral Nervous System Transmits Sensory Input

1. There are special senses and somatic senses.
2. The four classes of sensory receptors are: mechanoreceptors, chemoreceptors, thermoreceptors, and electroreceptors.

Mechanoreceptors Are Involved with Hearing and Balance

1. The hair cell in vertebrates is stimulated by mechanical pressure.
2. The hair cell receives sound waves for hearing. They detect vibrations in the cochlea of the ear. Sensory neurons communicate information to the brain.
3. Hair cells in the semicircular canals detect stimuli for orientation and gravity.

Photoreceptors Monitor Vision and Light Sensitivity

1. Visible light has a range of wavelengths that are interpreted as color.
2. The retina of the eye contains photoreceptors.
3. Cones are responsible for color vision.
4. Rods, with rhodopsin, are for vision in dim light.
5. The optic nerve sends signals from the eye to the brain.

Humans Have Color Vision

1. There are three different kinds of cones with opsins.

The Peripheral Nervous System: Motor Output

1. The somatic system sends motor signals to skeletal muscles.
2. The autonomic system sends motor signals to organs with smooth or cardiac muscle.
3. The autonomic system has two components: sympathetic and parasympathetic.

4. These two components have opposite effects on the activity of an internal organ. The sympathetic is dominant during fear or stress situations (i.e., increased rate of heartbeat).

STUDY HINTS

1. This chapter contains a wealth of information. Try to divide it into several separate units of study: central nervous system, peripheral nervous system, and receptors.
2. Continue to write some questions on your own for additional practice. Try a match on regions of the brain. List five events , in their correct order, that unfold when a neuron sends an impulse. Can you think of other useful exercises?

CHAPTER REVIEW

Key Terms

Define these key terms from the chapter:

1. neuron

2. cephalization

3. glial cell

4. sensory neuron

5. axon

6. action potential

7. synapse

8. neurotransmitter

9. autonomic system

10. corpus callosum

11. cortex

12. efferent

13. perception

14. retina

15. opsin

Chapter Concepts

1. How does the nervous system relate to the integration of all the organ systems you have now studied?
2. Return to the idea that structure determines function. How does this idea apply to the cerebrum of the human brain?

Practice Test

TRUE/FALSE

___ 1. The spinal cord is part of the peripheral nervous system.

___ 2. The neurons of the hydra are organized into a nerve net.

___ 3. Sensory neurons transmit signals toward the CNS.

___ 4. Within a neuron a signal normally travels from the axon to the cell body to the dendrites.

___ 5. The presence of the myelin sheath and nodes of Ranvier increases the ability of the neuron to regenerate.

___ 6. The extracellular concentration of sodium ions is higher than inside the neuron at rest.

___ 7. The resting potential of a neuron is restored after the action potential mainly by the flow of chloride ions out of the cell.

___ 8. The neurotransmitter is a chemical signal.

___ 9. A gas such as carbon monoxide can act as a neurotransmitter.

___ 10. The autonomic nervous system controls the motor function of skeletal muscles.

___ 11. The cerebrum is part of the hindbrain.

___ 12. In humans, the cerebrum makes up most of the brain's mass.

___ 13. The pons is part of the hindbrain.

___ 14. The cerebral cortex consists of outer white matter.

___ 15. Broca's area is located on the frontal lobe of the cerebral cortex.

___ 16. The primary visual cortex is located on the temporal lobe.

___ 17. The central sulcus is between the frontal and parietal lobes.

___ 18. Chemoreceptors detect stimuli for hearing and balance.

___ 19. The ossicles are located in the inner ear.

___ 20. Rhodopsin is a visual pigment found in the rods of the retina.

MULTIPLE CHOICE

21. In an animal, cephalization refers to the concentration of
    a. development of skeletal muscles and sensory organs.
    b. development of skeletal muscles and an endocrine system.
    c. sensory and integrative functions in the anterior end.
    d. sensory and integrative functions in the posterior end.
22. Glial cells
    a. contract to produce movement.
    b. destroy neurons.
    c. send signals.
    d. support and protect neurons.
23. Motor neurons send impulses
    a. away from muscles.
    b. toward effectors.
    c. toward interneurons.
    d. toward receptors.
24. At rest the electrical potential inside a neuron is
    a. negative.

b. neutral.
c. nonexistent.
d. positive.

25. In a neuron, a resting potential is converted to an action potential by the transport of __________ ions into the cell.
a. chloride
b. iodide
c. potassium
d. sodium

26. The speed of impulse transmission increases along a neuron with the
a. absence of myelin and nodes.
b. absence of myelin and presence of nodes.
c. presence of myelin and absence of nodes.
d. presence of myelin and nodes.

27. A neurotransmitter is released from the __________ of a neuron.
a. axon
b. cell body
c. dendrites
d. perikaryon

28. L-dopa is used to treat __________ disease.
a. Alzheimer's
b. Huntington
c. Parkinson's
d. protonopia

29. The autonomic nervous system signals each of the following except the
a. cardiac muscle.
b. skeletal muscles.
c. the stomach.
d. the uterus.

30. The diencephalon is part of the
a. brain stem.
b. cerebellum.
c. forebrain.
d. spinal cord.

31. The origin of most of the cranial nerves is from the
a. brain stem.
b. cerebellum.
c. cerebrum.
d. hypothalamus.

32. The hypothalamus controls each of the following except the
a. intake of food.
b. maintenance of balance.

c. output of urine.
d. thirst.

33. The corpus callosum
    a. connects the two cerebral hemispheres.
    b. controls body temperature.
    c. houses a sleep center.
    d. maintains body equilibrium.
34. Which brain region is an important link between the nervous system and endocrine system?
    a. hypothalamus
    b. medulla
    c. pons
    d. thalamus
35. The primary sensory cortex is found on the ___________ lobe.
    a. frontal
    b. occipital
    c. parietal
    d. temporal
36. Damage to Wernicke's area on the cerebral cortex impairs
    a. comprehension of words.
    b. control of blood pressure and respiration.
    c. equilibrium of the body.
    d. speech.
37. The dorsal and ventral roots are parts of the
    a. cerebral cortex.
    b. gray matter inside the spinal cord.
    c. skeletal muscle.
    d. spinal nerve.
38. Mechanoreceptors detect stimuli for
    a. electrical currents.
    b. hearing and balance.
    c. light.
    d. temperature.
39. The organ of Corti is found in the ___________ of the ear.
    a. auricle
    b. cochlea
    c. semicircular canals
    d. saccule and utricle
40. The function of the sclera of the eye is
    a. anchorage for extrinsic eye muscles.
    b. for detection of light.

c. for regulation of the shape of the lens.
d. for transmitting signals to the brain.

COMPLETION

41. Sound waves travel from the external auditory meatus to the ___________ membrane or eardrum.
42. When stimulated the hair cells in the cochlear duct rub against an overhanging ___________ membrane.
43. The ___________ regulates the amount of light passing through the pupil as it passes through the eye.
44. The ___________ humor fills the anterior chamber of the eye.

MATCHING

Label each one of the follow statements as describing the autonomic nervous system or somatic nervous system.

45. Acetylcholine is the only neurotransmitter: ___________ .
46. It has a two-neuron chain from its origin in the CNS to the effector organ: ___________ .
47. It has sympathetic and parasympathetic terminals: ___________ .
48. Two branches are present to stimulate and inhibit the activity of an organ: ___________ .
49. It controls only skeletal muscles: ___________ .
50. It is under involuntary control but is subject to voluntary control through biofeedback: ___________ .

ANSWERS

Key Terms

1. The neuron is the nerve cell that sends signals.
2. Cephalization is the concentration of sensory and integrative functions in the anterior end of the animal.
3. The glial cell protects and supports the neuron.
4. A sensory neuron sends a signal from a receptor to the central nervous system.

5. The axon is the last part of the neuron to send an impulse.
6. The action potential is the electrical impulse that travels along the neuron when it is sending a signal.
7. The synapse is the junction between neurons.
8. The neurotransmitter is the chemical signal at the synapse.
9. The autonomic system is the part of the peripheral nervous system that sends motor signals to organs with smooth and cardiac muscle.
10. The corpus callosum is a bridge connecting the two hemispheres of the cerebrum.
11. The cortex is the outer gray layer of each cerebral hemisphere.
12. Efferent means motor, sending a signal away from the central nervous system.
13. Perception is the interpretation of the external environment.
14. The retina is the inner sensory layer of the eye.
15. An opsin is one of three different pigments in the cones of the retina.

Chapter Concepts

1. The nervous system establishes a great telegraph system throughout the body. When a person runs, for example, this system signals the heart, diaphragm, digestive tract, skeletal muscles, and many other structures. All of them must work together (integration).
2. The cerebral cortex is convoluted or wrinkled. This increases its surface area and increases sensory, motor, and associative ability from the organism. The associative areas determine abstract abilities such as thinking and memory.

Practice Test

1. False
2. True
3. True
4. False
5. False
6. True
7. False
8. True
9. True
10. False
11. False
12. True
13. True
14. False
15. True
16. False
17. True
18. False
19. False
20. True
21. c
22. d
23. b
24. a
25. d
26. d
27. a
28. c
29. c
30. c
31. a
32. b
33. a
34. a
35. c
36. a
37. d
38. b
39. b
40. a
41. tympanic
42. tectorial
44. aqueous
45. somatic
46. autonomic
47. autonomic
48. autonomic
49. somatic
50. autonomic

❧ Chapter 26 ☙

THE ENDOCRINE SYSTEM

CHAPTER OVERVIEW

This chapter presents the control system that works in conjuction with the nervous system, the endocrine system. The mechanism of endocrine control, through the release of hormones, is different, however. A series of glands secrete chemical messengers (hormones) that are transported in the blood. They control a wide array of body processes ranging from bone growth to the concentration of glucose in the blood.

As a control system the endocrine system is a major contributor to homeostasis. Stimuli such as temperature changes and dietary patterns require responses to maintain a constant internal environment.

LEARNING OBJECTIVES

After studying this chapter, you should be able to

1. Outline the mechanism of endocrine control.
2. Compare the several kinds of hormones chemically.
3. Explain the mode of hormone action at the cellular level.
4. Explain the relationship between the nervous and endocrine systems.
5. List the major endocrine glands and their hormones.
6. Describe the action of each hormone.
7. State some results in body function from an imbalance of hormones.

CHAPTER OUTLINE

The Endocrine System Coordinates Long-Term Responses to Stimuli

1. It works with the nervous system for homeostasis.

2. It consists of a series of endocrine glands.
3. These glands send hormones that control long-term responses.

Endocrine Systems Have Evolved Several Times

1. They are found in crustaceans, arthropods, and vertebrates.
2. In vertebrates the system consists of about one dozen glands.
3. The glands develop from the three primary germ layers.

Endocrine Systems Use Feedback Systems and Cycles for Regulation

1. The regulation is through negative feedback.
2. The release of TSH to control the thyroid hormone is one example.
3. The system controls cycles of several durations.

Hormones Belong to Three Chemical Classes

1. Steroids include testosterone and estradiol.
2. Peptides are secreted from a variety of organs.
3. Amines are derived from tyrosine.

How Hormones Are Synthesized, Stored, and Secreted

1. Steroid hormones are derived from cholesterol. They directly enter the bloodstream.
2. Peptide hormones enter the blood from storage sites that make them.
3. Epinephrine is a major amine hormone.

Hormones Act on Target Cells by Two Mechanisms

1. A hormone binds to a receptor on the target cell. This changes its shape and activity.
2. A nonsteroid hormone binds to receptors on the plasma membrane. It activates a secondary messenger that produces a target cell response.
3. A steroid hormone passes through the plasma membrane of the target cell and binds to the nucleus. This activates genes for transcription.
4. Problems with hormones can be production imbalances and nonfunctional receptors.

The Hypothalamus-Pituitary Axis Connects the Nervous and Endocrine Systems

1. The anterior lobe of the pituitary is glandular tissue.
2. The posterior lobe is nervous tissue.
3. The hypothalamus controls the release of hormones from these lobes.

Anterior Pituitary Hormone Release Is Controlled by Neurosecretion

1. Hypothalmic hormones reach the anterior pituitary through a portal system.

2. Hypothalmic hormones control the release of the anterior pituitary hormones.

Growth Hormone Stimulates Bone Growth
1. It is released by a hypothalmic hormone and inhibited by another hormone.
2. GH stimulates protein synthesis in target cells.
3. It can promote mineral growth in bones.
4. Imbalances include dwarfism and gigantism.

Thyroid-Stimulating Hormone Regulates the Thyroid
1. The thyroid hormones regulate metabolic rates and body temperature.

The Anterior Pituitary Secretes Hormones Involved in Reproduction
1. The gonadotropins are FSH and LH.
2. Prolactin affects milk production.
3. ACTH signals the adrenal glands.

The Posterior Pituitary Is an Extension of the Brain
1. ADH signals the kidneys to reabsorb water.
2. Oxytocin stimulates the contraction of the smooth muscle of the uterus.

Other Endocrine Organs Have Specialized Functions

The Adrenal Glands Regulate Responses to Stress
1. Each gland consists of the cortex and medulla.
2. The adrenal medulla secretes epinephrine and norepinephrine. These hormones have effects that match the effects of the sympathetic nervous system.
3. The adrenal cortex secretes mineralcorticoids (mineral balance), glucorticoids (stress, inflammatory response, metabolism of molecules), and sex hormones.

The Thyroid Gland Helps Regulate Metabolism and Growth
1. The follicles store a precursor to the thyroid hormone.
2. TSH stimulates the release of the thyroid hormone.
3. This hormone increases the metabolic rate.
4. Calcitonin increaes the excretion of calcium from the kidneys.

The Pancreas Regulates Glucose Metabolism
1. It consists of pancreatic islets.
2. Insulin removes glucose from the blood if it is too high. It is stored.
3. Glucagon releases glucose into the blood from storage.
4. Type I diabetes results from a deficiency of insulin. There are disturbances in the metabolism of many molecules.

5. The symptoms of type II diabetes are different and require different treatment.

Other Chemical Messengers Share Characteristics of Hormones

Interferons Protect against Viral Infections

1. They are proteins and can also inhibit cancer cell growth.

Prostaglandins Are Local-Acting Signals

1. They act on nearly all cells with diverse effects.

Pheromones Act as Signals between Organisms

1. They function in social communication.

The Endocrine System Regulates Many Biological Cycles

Circadian Rhythms Are Daily Cycles

1. Cyclic hormones control them.

Many Biological Processes Are Regulated by Rhythms

1. The menstrual cycle and childbirth are examples.

Hypothalamic Mechanisms Control Cycles and Rhythms

1. The SCN sends a signal to the pineal gland.
2. Pituitary secretions regulate the cycle.

STUDY HINTS

1. Make a table listing the endocrine glands in the lefthand column. In two columns to the right show the hormones from each gland and the function of each hormone.
2. Continue to study the key terms as well as the questions and problems before and after studying the chapter.

CHAPTER REVIEW

Key Terms

Define these key terms from the chapter:

1. hormone

2. negative feedback

3. peptide

4. second messenger

5. portal system

6. vasopressin

7. cortex

8. follicle

9. pheromone

10. circadian

Chapter Concepts

1. What does negative feedback mean as a response pattern for homeostasis? Does "negative" mean there is always a decrease?
2. Does endocrine control duplicate some of the examples of control you have already studied? What are they?

Practice Test

TRUE/FALSE

___ 1. Endocrine glands secrete chemical messengers called hormones.

___ 2. TSH is secreted by the thyroid gland.

___ 3. Some hormones are steroid molecules.

___ 4. Testosterone is a steroid.

___ 5. Epinephrine is not an amine hormone.

___ 6. Cyclic AMP is a second messenger.

___ 7. There is a portal system between the thalamus and hypothalamus of the brain.

___ 8. GH is secreted by the anterior lobe of the pituitary gland.

___ 9. The growth hormone stimulates protein synthesis in cells.

___ 10. Dwarfism and gigantism develop from disorders of the thyroid gland.

___ 11. ACTH affects the activity of the adrenal gland.

___ 12. Oxytocin is a steroid hormone.

___ 13. The release of epinephrine is stimulated by the parasympathetic nervous system.

___ 14. The adrenal cortex secretes cortisol.

___ 15. Calcitonin stimulates the retention of calcium in the blood by signaling the kidneys.

___ 16. Insulin acts to lower the concentration of glucose in the blood.

___ 17. Type II diabetes is associated with abnormally low levels of glucose in the blood.

___ 18. Interferons are released in response to viral infections.

___ 19. Prostaglandins are carbohydrates.

___ 20. Pheromones are chemical signals secreted into the blood.

MULTIPLE CHOICE

21. Which of the following glands is not an example of an exocrine gland?
    a. digestive
    b. endocrine
    c. salivary
    d. sweat
22. Which of the following statements is not a characteristic of the endocrine system?
    a. It is dependent on specificity of target.
    b. It functions over long distances in the body.
    c. It releases hormones into the blood.
    d. It can signal rapid responses within milliseconds.
23. Which of the following choices is not a chemical family of hormones?
    a. amines
    b. nucleic acids

c. peptides
d. steroids

24. Amines are derived from the amino acid
a. lysine.
b. phenylalanine.
c. tryptophane.
d. tyrosine.

25. Select the correct statement about steroid hormones.
a. They are derived from cholesterol.
b. They are mostly bound to lipids in the blood.
c. They are soluble in water.
d. They are stored in large amounts.

26. Select the steroid hormone.
a. estradiol
b. epinephrine
c. insulin
d. mRNA

27. Which of the following is not a problem of the endocrine system?
a. functional receptors of target cells
b. nonfunctional receptors of target cells
c. overproduction of endocrine cells
d. underproduction of endocrine cells

28. The pituitary gland is physically linked to the ___________ of the brain.
a. cerebellum
b. hypothalamus
c. medulla
d. pons

29. A portal system begins and ends in
a. capillaries.
b. glial cells.
c. lymph cells.
d. neurons.

30. Select the hormone secreted by the anterior pituitary gland.
a. epinephrine
b. FSH
c. glucagon
d. TH

31. The hyperplasia of cells means that they
a. change location in the body.
b. decrease in number.
c. increase in number.
d. stay in one location in the body.

32. Select the hormone secreted by the posterior pituitary gland.
    a. ACTH
    b. ADH
    c. FSH
    d. LH
33. One function of oxytocin is to
    a. decrease the metabolic rate.
    b. decrease the concentration of calcium in the blood.
    c. stimulate uterine contraction.
    d. stimulate water reabsorption in the kidney.
34. The cortex and medulla are layers of the __________ gland.
    a. adrenal
    b. hypothalmic
    c. pituitary
    d. thyroid
35. Another name for __________ is vasopressin.
    a. ACTH
    b. ADH
    c. TH
    d. TSH
36. Which of the following choices is a class of adrenal steroids?
    a. catecholamines
    b. glucocorticoids
    c. mineralcorticoids
    d. sex hormones
37. Glucagon opposes the action of
    a. ADH.
    b. calcium.
    c. glycogen.
    d. insulin.
38. A circadian rhythm occurs every __________ hours.
    a. 2
    b. 12
    c. 24
    d. 168
39. Diabetes mellitus, type I, is associated with a deficiency of
    a. glucagon.
    b. insulin.
    d. lipids.
    d. MSH.

40. The ___________ gland is located between the two cerebral hemispheres.
    a. adrenal
    b. pineal
    c. pancreatic
    d. thyroid

COMPLETION

41. Endocrine systems are found in vertebrates, crustaceans, and ___________ .

42. Amines are soluble in ___________ .

43. ___________ are short chains of amino acids.

44. GMP is a ___________ messenger, as it works inside a target cell.

45. The posterior lobe of the pituitary gland is composed of ___________ tissue.

46. GH stimulates ___________ , cells that deposit the mineral part of bone tissue.

47. LH stimutates ___________ formation and steroid hormone production.

48. Epinephrine reinforces the action of the ___________ nervous system for "fight or flight" responses.

49. The thyroid hormone ___________ the overall metabolic rate.

50. In type ___________ diabetes, cells do not respond to insulin.

ANSWERS

Key Terms

1. A hormone is a chemical messenger secreted into the blood by an endocrine gland.
2. Negative feedback involves a response that opposes the direction of the original response.
3. A peptide is a short chain of amino acids that composes most hormones.
4. A second messenger is generated in a cell that is signaled by a nonsteroid hormone.
5. A portal system ranges from one group of capillaries to a second group of capillaries.
6. Vasopressin is another name for ADH, secreted by the posterior lobe of the pituitary.

7. The cortex is the outer layer of the adrenal gland.
8. The follicle is a spherical structure in the thyroid gland.
9. A pheromone is a chemical signal that travels between organisms.
10. A circadian rhythm fluctuates about every 24 hours.

Chapter Concepts

1. The key meaning for negative is *reverse.* If body temperature rises, the body responds to lower it. If the level of sugar in the blood decreases, the body responds to raise it. In either direction some optimal level is maintained.
2. The best examples include a comparison of epinephrine/norepinephrine actions and the sympathetic nervous system. Refer to what you have learned about the motor control of the internal organs.

Practice Test

1. True	14. True	27. a	40. b
2. False	15. False	28. b	41. arthropods
3. True	16. True	29. a	42. water
4. True	17. False	30. b	43. peptides
5. False	18. True	31. c	44. second
6. True	19. False	32. b	45. nervous
7. False	20. False	33. c	46. osteoblasts
8. True	21. b	34. a	47. gamete
9. True	22. d	35. b	48. sympathetic
10. False	23. b	36. a	49. increases
11. True	24. d	37. d	50. II (two)
12. False	25. a	38. c	
13. False	26. a	39. b	

Chapter 27

PLANT REPRODUCTION AND DEVELOPMENT

CHAPTER OVERVIEW

This chapter describes the strategies that plants have evolved to reproduce and develop successfully. The angiosperms are the dominant land plants. Gamete production occurs in the flower. The subsequent events of reproduction and development begin there. The entire life cycle of the flowering plant involves the alternation of a diploid sporophyte with a haploid gametophyte. Pollination, plus seed and fruit production, are strategies for plant success on the Earth.

Plant growth is regulated by a variety of chemical signals. This is a topic of botany with many unanswered questions to be solved by future research.

LEARNING OBJECTIVES

After studying this chapter, you should be able to

1. Outline the life cycle of an angiosperm.
2. Describe the structures and functions of the flower.
3. Explain meiosis and double fertilization in the angiosperm.
4. Explain the advantages of seed and fruit production.
5. Outline the pattern of plant development.
6. List some major plant hormones with accompanying functions.
7. State some examples of vegetative reproduction and cloning in plants.

CHAPTER OUTLINE

Reproduction in Plants: A Preview

1. Angiosperms are the dominant land plants.

2. Their life cycle is an alternation of generations: diploid sporophyte with a haploid gametophyte.
3. Flowers promote the spread of pollen. Fruits promote seed dispersal.
4. The flowering plants can reproduce sexually and asexually.

Gamete Formation Takes Place in Flowers

1. Sexual reproduction requires gamete production and fusion.

Flowers: Structure and Function

1. The flower is a complex organ on a modified shoot.
2. It consists of sepals and petals.
3. The stamen is the male reproductive structure. Anthers produce pollen grains, the male gametophytes.
4. Carpels compose the female reproductive structure. A carpel consists of the ovary, style, and stigma.
5. The female gametophyte develops in the ovule.

Meiosis Produces Gametophytes, Not Gametes

1. It occurs in anthers and ovules.
2. Gametes fuse to form a zygote that develops into a diploid sporophyte.
3. The spores are microspores and megaspores.
4. The endosperm provides nutrition for a developing embryo.

Pollination Is Followed by Fertilization

1. Pollen grains are dispersed in several ways.

Flowering Plants Employ Double Fertilization

1. Two sperm nucleii unite with a central cell producing a 3n cell which becomes the endosperm.
2. Another fertilzation forms a diploid embryo.

Plants and Animals Use Different Reproductive Strategies

1. In plants gametes are produced from active body cells in the plant.
2. Embryonic development establishes a basic body plan.

Seeds and Fruits Aid in Dispersal and Protect the Plant Embryo

1. The parent protects and nourishes the embryo.

The Development and Evolutionary Advantages of Seeds

1. The seed contains an embryo that is dormant for some time.
2. The endosperm supplies nutrition.
3. Seeds can tolerate stress and promote dispersal.

Fruits Protect Seeds and Promote Seed Dispersal

1. Fruits may be fleshy (soft or hard) or or dry.
2. Fruit types include simple, aggregate, multiple, and accessory.

Patterns of Growth in Embryos and Seedlings

1. Germination begins when the embyro breaks through the seed.
2. Root systems start to grow.
3. Cotyledons are seed leaves. They begin photosynthesis when the shoot emerges through the soil.
4. Plant development is an orderly series of events.

Plant Growth Regulators Control Development

1. Chemical messengers control development.

The Five Known Types of Plant Hormones

1. Auxins stimulate the elongation of the plant tips.
2. Gibberellins stimulate normal shoot elongation.
3. Cytokinins promote cell division and inhibit the aging of green tissues.
4. Abscisic acid promotes dormancy in perennial plants.
5. Ethylene stimulates fruit ripening.

Other Substances Act as Plant Hormones

1. Florigen may trigger flowering.

Vegetative Reproduction and Cloning

1. They can reproduce from runners.
2. Rhizomes can make new plants.
3. Roots can reproduce asexually.
4. These are adaptations.

Plants Can Be Grown from Single Cells

1. Studies of callus growth show this.
2. Genetically identical trees can be grown.
3. Synthetic seeds can be produced.

STUDY HINTS

1. Outline the life cycle of an angiosperm on a blank sheet of paper after studying the chapter. Include key labels such as diploid, haploid, sporophyte, and gametophyte.
2. List some major plant hormones. From memory explain the biological roles of each.

CHAPTER REVIEW

Key Terms

Define these key terms from the chapter:

1. angiosperm
2. sepal
3. petal
4. stamen
5. carpel
6. microspore
7. megaspore
8. endosperm
9. double fertilization
10. auxin

Chapter Concepts

1. Does a diploid organism, which is dominant among angiosperms, have a selective advantage over a haploid stage?
2. Does sexual reproduction, carried out by the flowering plants, have an advantage over asexual reproduction?

Practice Test

TRUE/FALSE

___ 1. Angiosperrms are the dominant plant on land on the Earth.

___ 2. Flowers form on the gametophyte generation of an angiosperm.

___ 3. Some angiosperms can reproduce asexually.

___ 4. Sepals are the modified leaves on a flower.

___ 5. The stamens are the female reproductive part of a flower.

___ 6. The filament of a stamen is topped by an anther.

___ 7. Pollen grains are the female gametophyte of the flower.

___ 8. Ovules of the flower contain cells that form the male gametophyte.

___ 9. Imperfect flowers contain male and female reproductive parts.

___ 10. Meiosis occurs in the anthers of flowers.

___ 11. Meiosis does not occur in the ovules of flowers.

___ 12. Microsprores are haploid cells.

___ 13. An ovule differentiates into a seed after fertilization.

___ 14. In plants, embryonic development forms a complete adult in minature.

___ 15. True fruits develop from an ovary.

___ 16. Wind can disperse seeds over long distance.

___ 17. Auxins stimulate the growth of lateral buds.

___ 18. Rhizomes are underground stems.

___ 19. Ethylene gas is not a plant hormone.

___ 20. Plants can be grown from single cells.

MULTIPLE CHOICE

21. In angiosperms, there has been the evolution of
    a. flowers only.
    b. fruits only.
    c. flowers and fruits.
    d. neither flowers nor fruits.

22. Among angiosperms
    a. diploid spores develop into gametophytes.
    b. diploid spores develop into sporophytes.
    c. haploid spores develop into gametophytes.
    d. haploid spores develop into sporophytes.
23. In angiosperms, megaspores form the
    a. female gametophyte.
    b. female sporophyte.
    c. male gametophyte.
    d. male sporophyte.
24. Bud sports are known to produce a unique kind of
    a. fruit.
    b. leaf.
    c. rhizome.
    d. root.
25. The outermost parts of the flower are
    a. anthers and filaments.
    b. ovules and stamens.
    c. petals and sepals.
    d. petals and stamens.
26. Insects are mainly attracted to flowers by
    a. color.
    b. odor.
    c. position.
    d. sound.
27. Birds are mainly attracted to flowers by
    a. color.
    b. odor.
    c. position.
    d. sound.
28. Which of the following choices is not a male reproductive structure?
    a. anther
    b. filament
    c. pollen grain
    d. style
29. Select the structure that is a combination of the other three choices.
    a. carpel
    b. ovary
    c. stigma
    d. style

30. The top part of the female reproductive structure of the flower is the
    a. ovary.
    b. ovule.
    c. stigma.
    d. style.
31. A perfect flower has
    a. an absence of reproductive parts.
    b. male and female reproductive structures.
    c. only female reproductive structures.
    d. only male reproductive structures.
32. Meiosis occurs in
    a. anthers and ovules.
    b. anthers and petals.
    c. petals and sepals.
    d. sepals and rhizomes.
33. The male gametophyte forms in the
    a. anther.
    b. ovule.
    c. petal.
    d. sepal.
34. The female gametophyte forms in the
    a. anthers.
    b. ovules.
    d. petals.
    d. sepals.
35. The function of the endosperm is to
    a. anchor the plant.
    b. conduct meiosis.
    c. obtain water.
    d. supply nutrients.
36. Double fertilization in plants forms a ___________ cell.
    a. n
    b. 2n
    c. 3n
    d. 4n
37. A cotyledon is a
    a. developed flower.
    b. modified stem.
    c. seed leaf.
    d. specialized root system.

38. Gibberellins
    a. inhibit plant dormancy.
    b. inhibit the elongation of the shoot.
    c. stimulate plant dormancy.
    d. stimulate the elongation of the shoot.
39. Abscisic acid is produced mainly in
    a. flowers.
    b. leaves.
    c. roots.
    d. stems.
40. Select the hormone that is mainly known to stimulate fruit ripening.
    a. auxin
    b. cytokinin
    c. ethylene
    d. gibberellin

COMPLETION

41. The gametophyte of angiosperms is ___________ , as it contains only one set of chromosomes.

42. The anther and filament compose the ___________ of the flower.

43. The ___________ is a tube formed in the ovary wall of the flower.

44. The number of microspores formed from each cell in the pollen sac is ___________ .

45. The ovary of a flower differentiates into a ___________ .

46. ___________ is a plant hormone that causes the root and shoot to elongate.

MATCHING

Each choice is used once.

___	47. apple	a. accessory fruit
___	48. grape	b. aggregate fruit
___	49. pineapple	c. multiple fruit
___	50. raspberry	d. simple fruit

ANSWERS

Key Terms

1. The angiosperm is the flowering plant.
2. Sepals are the green modified leaves that are outermost for the flower.
3. Petals are colored structures of the flower.
4. The stamen is the male reproductive structure of the flower.
5. The carpel is the female reproductive structure of the flower.
6. The microspore is the haploid structure forming a pollen grain.
7. The megaspore is the haploid structure forming a female gamete.
8. The endosperm provides nutrition for the plant embryo.
9. Through double fertilization, two sperm nucleii fertilize a central cell and another nucleus fertilizes a haploid egg.
10. Auxin is a family of hormones (auxins) that stimulate elongation of the tips of a growing plant.

Chapter Concepts

1. Return to your study of genetics. Diploidy allows a dominant allele to mask a recessive allele in an organism. Meiosis produces a variety of sex cells. Genetic variability is enhanced.
2. Sexual reproduction promotes genetic variability. A population with variation has diversity toward natural selection.

Practice Test

1. True	14. False	27. a	40. c
2. False	15. True	28. d	41. haploid
3. True	16. True	29. a	42. stamen
4. True	17. False	30. c	43. style
5. False	18. True	31. b	44. four
6. True	19. False	32. a	45. fruit
7. False	20. True	33. a	46. auxin
8. False	21. c	34. b	47. a
9. False	22. c	35. d	48. d
10. True	23. a	36. c	49. c
11. False	24. a	37. c	50. b
12. True	25. c	38. d	
13. True	26. b	39. b	

Chapter 28

ANIMAL REPRODUCTION AND DEVELOPMENT

CHAPTER OVERVIEW

This chapter begins by explaining some examples of asexual reproduction among animals. The general stages of prenatal development are shown next. Several factors influence this.

Most of the information centers on the human reproductive systems. The structures and functions of each sex are described. The female reproductive cycle is an excellent example of a process controlled by the fluctuating levels of several hormones. The understanding of STDs and reproductive technology for both sexes continues to increase.

Much is known about the developmental stages of human prenatal development. Cells specialize to form complex tissues and organs. Events are listed over three trimesters. Birth is hormonally induced.

LEARNING OBJECTIVES

After studying this chapter, you should be able to

1. Compare several means of asexual reproduction in animals.
2. Outline the general steps of development before birth.
3. Discuss some factors affecting development before birth.
4. Describe the structures and functions of the human male reproductive system.
5. Describe the structures and functions of the female reproductive system.
6. Explain how changing levels of hormones regulate the female reproductive cycles.

7. Outline the steps of human sexual response.
8. List some STDs.
9. Discuss some advances in reproductive technology.
10. List some events in each trimester of development before birth.
11. List some events associated at birth.
12. Outline the stages of postnatal development.

CHAPTER OUTLINE

Principles of Animal Development

Asexual Reproduction Produces Offspring from a Single Parent

1. The offspring are genetically identical.
2. Some invertebrates undergo budding.
3. Sea stars reproduce by fragmentation.
4. Asexual reproduction has some adaptations.

Sexual Reproduction Increases Genetic Variability among Offspring

1. Gametes unite to form a zygote.
2. Some organisms reproduce sexually and asexually.
3. Sexual reproduction has advantages and costs.

Fertilization and Cleavage Are Initial Stages of Development

1. Gametes fuse during fertilization.
2. Fertilization transmits genes, restores the diploid number, and initiates development.
3. The cleavage of a zygote produces a blastula.
4. Important molecules are distributed to the embryo.

Gastrulation Produces an Organized, Three-Layered Embryo

1. The three primary layers are the ectoderm, mesoderm, and endoderm.
2. Each layer develops into specialized tissues and organs.
3. The ectoderm forms a neural tube. The mesoderm forms a notochord.
4. The mesoderm forms somites.

Pattern Formation and Induction Result in Organ Formation

Pattern Formation Lays Out the Body Plan

1. A homeobox responds to regulatory molecules to establish a body plan.

Induction Starts Organ Formation

1. One cell or tissue affects the development of another cell or tissue.
2. It affects the pattern of gene expression in cells.
3. One example is the formation of the vertebrate eye.

Human Reproduction and Development

1. Fertilization is internal and controlled by the hypothalamus and endocrine system.
2. The system in each sex consists of gonads with accessory glands and ducts.

Anatomy of the Male Reproductive System

1. The testis can descend into the scrotum, depending on temperature.
2. The seminiferous tubules produce sperm cells.
3. Sperm cells mature as they are stored in the epididymis.
4. FSH controls sperm cell production.
5. LH controls testosterone production in the testis.
6. Migrating sperm cells travel through a series of ducts: vas deferens, ejaculatory duct, and urethra.
7. Along this route glands add seminal fluid in the order: seminal vesicles, prostate gland, and bulbourethral gland.

Anatomy of the Female Reproductive System

1. The ovary contains follicles. Each follicle surrounds a primary oocyte.
2. A developing oocyte is released during ovulation.
3. The ovaries cycle between a follicular phase and luteal phase.
4. About midway in the ovarian cycle ovulation occurs. The follicle is converted into a corpus luteum.
5. Fertilization usually occurs in the oviduct.
6. The endometrium is the inner lining of the uterus. The cervix of the uterus fits into the vagina.
7. The vulva are the external genitalia.

Hormones Regulate the Ovarian Cycle and the Menstrual Cycle

1. The levels of estrogen and progesterone fluctuate.

The Menstrual Cycle Is Synchronized with the Ovarian Cycle

1. The length is about 28 days.
2. It begins with menstruation.
3. FSH increases and signals a follicle to mature.
4. LH peaks to stimulation ovulation.
5. The corpus luteum secretes more progesterone and estrogen.
6. These ovarian hormones signal development of the endometrium for possible implantation of an embryo.
7. If fertilization and implantation do not occur the endometrium is expelled.

Human Sexual Responses

1. Blood flows through three cylinders of spongy tissue in the penis.

2. There are erections in both sexes.
3. The female sexual response phases are: arousal, plateau, orgasm, and resolution.

Sexually Transmitted Diseases

1. The most common producing inflammation are gonorrhea and chlamydia.
2. A second class produces sores on the external genitalia.
3. A third class is viral diseases that affect organ systems.

Reproduction and Technology

Contraception Uncouples Sexual Intercourse from Pregnancy

1. Methods include abstinence, tubal ligation, vasectomy, and birth control pills.
2. Chemical and physical barriers can prevent fertilization.

Reproductive Technology Can Enhance Fertility

1. Infertility can be related to problems in the reproductive tract.
2. *In vitro* fertilization offers one solution.
3. Reproductive technology has offered new solutions.

Human Development from Fertilization to Birth

1. Fertilization usually occurs in the upper third of the oviduct.
2. Various mechanisms promote this.
3. A sperm fertilizes an oocyte.
4. The zygote forms an embryo.
5. The morula converts into a blastula.
6. Trophoblast cells implant onto the endometrium.
7. The trophoblast forms a chorion.
8. hCG prolongs the life of the corpus luteum.
9. Villi establish a maternal-embryo circulation.
10. A placenta with umbilical cord is established.

Human Development Is Divided into Three Stages

1. They are three trimesters.

Rapid and Complex Events Occur in the First Trimester

1. The changes are rapid.
2. There is a lengthening of the embryo. A four-chambered heart and limb buds are formed in the second month.
3. Chromosomal sex is determined in week 7.
4. Male and female gonads are formed under hormone signaling.

The Second Trimester Is a Period of Organ Maturation

1. Organs become well-formed. There are other well-formed features.

The Last Trimester Is a Period of Rapid Growth

1. The fetus doubles in size in the last two months.
2. The baby's immune system will mature after birth.

Birth Is Hormonally Induced and Occurs in Stages

1. The responses are under positive feedback.
2. Oxytocin stimulates uterine contraction.
3. The cervix dilates in the first stage.
4. There are strong uterine contractions in the second stage.
5. A second round of uterine contractions begins the third stage.

Milk Production Begins Shortly after Birth

1. Prolactin promotes milk production.

Postnatal Developmental Landmarks

Infancy and Childhood

1. Infancy is from birth through the first two years.
2. Childhood begins at three years age.

Adolescence and Adulthood

1. Growth is accelerated during puberty.
2. Organ systems grow greatly during adolescence.
3. Aging begins in the late twenties.
4. Cellular function and division diminishes.
5. Medical advances are increasing life spans.

STUDY HINTS

1. List the major structures of the human male and female reproductive systems. After studying the chapter write functions of these from memory.
2. State the key events in the female reproductive cycles (menstrual and ovarian). State each event on a separate card. Mix up the cards and put the events in the correct order for each cycle.

CHAPTER REVIEW

Key Terms

Define these key terms from the chapter:

1. asexual reproduction

2. zygote

3. blastula

4. mesoderm

5. somite

6. induction

7. gona

8. follicle

9. luteal phase

10. menstruation

11. vasectomy

12. fertilization

13. trophoblast

14. placenta

15. villi

Chapter Concepts

1. Science is an ongoing process, offering some answers with new questions. What are some unanswered questions about development?
2. Explain some earlier concepts of physiology that help you to understand the female reproductive cycles.

Practice Test

TRUE/FALSE

___ 1. Fragmentation is one form of sexual reproduction.

___ 2. The blastula is a ball of cells produced from the cleavage of the zygote.

___ 3. Somites of the embryo produce the muscles of the body wall.

___ 4. Induction affects the activation and inactivation of genes during development.

___ 5. The seminiferous tubules produce testosterone.

___ 6. The epididymis stores sperm.

___ 7. The follicular phase is associated with the corpus luteum.

___ 8. Cilary action moves the secondary oocyte through the oviduct.

___ 9. LH is secreted by the posterior lobe of the pituitary gland.

___ 10. FSH stimulates the follicle to secrete estrogen.

___ 11. Chlamydia is caused by a yeast infection.

___ 12. Tubal ligation is the cutting off and sealing off of the oviduct.

___ 13. *In vitro* fertilization occurs inside the body.

___ 14. The oviduct is the most common site of fertilization.

___ 15. Implantation is the attachment of the trophoblast to the endometrium.

___ 16. The hormone hCG is the one detected in the pregnancy test.

___ 17. The zygote divides by meiosis.

___ 18. In humans XY fertilizations produce males.

___ 19. DHT in humans prevents the development of male genitalia.

___ 20. The release of oxytocin inhibits uterine contractions.

MULTIPLE CHOICE

21. Which of the following choices is not a type of asexual reproduction?
    a. budding
    b. fission
    c. fragmentation of the organism
    d. fusion of gametes
22. The early stages of development occur in the order
    a. blastula, gastrula, zygote.
    b. blastula, zygote, gastrula.
    c. zygote, blastula, gastrula.
    d. zygote, gastrula, blastula.
23. Select the structure that develops from the mesoderm.
    a. bone
    b. brain
    c. epidermis
    d. gallbladder
24. Select the structure that develops from the endoderm.
    a. blood vessels
    b. dermis
    c. inner ear
    d. liver
25. After the epididymis, migrating sperm cells next move into the
    a. ejaculatory duct.
    b. testis.
    c. urethra.
    d. vas deferens.
26. Select the correct order of glands by which seminal fluid is added to the male reproductive tract.
    a. bulbourethral glands, prostate gland, seminal vesicles
    b. bulbourethral glands, seminal vesicles, prostate gland
    c. seminal vesicles, bulbourethral glands, prostate gland
    d. seminal vesicles, prostate gland, bulbourethral glands
27. Select the diploid structure.
    a. ovum
    b. polar body
    c. primary oocyte
    d. secondary oocyte
28. When does ovulation normally occur in the ovarian cycle?
    a. beginning
    b. end
    c. midpoint
    d. near the end

29. The ____________ is the middle layer of the uterus.
    a. cervix
    b. endometrium
    c. epimetrium
    d. myometrium
30. Which hormone reaches a peak concentration in the blood during ovulation?
    a. estradiol
    b. FSH
    c. LH
    d. progesterone
31. Gonorrhea is caused by a
    a. bacterium.
    b. fungus.
    c. protozoan.
    d. virus.
32. Genital herpes is caused by a
    a. bacterium.
    b. fungus.
    c. protozoan.
    d. virus.
33. The chorion consists of ____________ layers.
    a. two
    b. three
    c. four
    d. five
34. Villi are
    a. pockets of the placenta.
    b. projections of the chorion.
    c. segments of the somites.
    d. vessels in the umbilical cord.
35. MIH causes the
    a. degeneration of the female chromosomes.
    b. degeneration of the female duct system.
    c. development of the female chromosomes.
    d. development of the female duct system.
36. Fetal movements normally begin in the ____________ month of pregnancy.
    a. third
    b. fifth
    c. sixth
    d. seventh

37. By the end of the third month of pregnancy the fetus is about ___________ cm long.
    a. three
    b. five
    c. nine
    d. eighteen
38. The fetus is expelled mainly by
    a. contractions of the cervix and dilation of the uterus.
    b. contractions of the uterus and dilation of the cervix.
    c. contractions of the cervix and uterus.
    d. dilation of the cervix and uterus.
39. ___________ stimulate uterine contractions.
    a. amines
    b. carbohydrates
    c. lipoproteins
    d. prostaglandins
40. Prolactin is released from the
    a. adrenal glands.
    b. anterior pituitary gland.
    c. posterior pituitary gland.
    d. thyroid gland.

MATCHING

Match each description, by letter, to the correct structure. Each choice is used once.

___ 41. oviduct	a. transports sperm cells to ejaculatory duct
___ 42. ovary	b. anterior lobe secretes FSH
___ 43. penis	c. receives the penis during intercourse
___ 44. pituitary	d. stores undeveloped female sex cells
___ 45. scrotum	e. conducts embryo to uterus
___ 46. testis	f. site of development of embryo and fetus
___ 47. urethra	g. has three cylinders of spongy tissue
___ 48. uterus	h. produces male sperm cells
___ 49. vagina	i. provides correct temperature for testis
___ 50. vas deferens	j. serves the reproductive and urinary tracts

ANSWERS

Key Terms

1. In asexual reproduction sex cells do not fuse.
2. The zygote is the fertilized egg.
3. The blastula is the hollow ball of cells produced from the cleavage of the zygote.
4. The mesoderm is the middle primary germ layer of the embryo.
5. A somite is a mesodermal structure that forms muscles in the body wall.
6. By induction a cell or tissue affects the developmental fate of another cell or tissue.
7. The gonad is the testis or ovary.
8. The follicle is the covering around each primary oocyte.
9. The luteal cycle occurs after ovulation.
10. Menstruation is the flow of blood in the female reproductive cycle.
11. A vasectomy is the cutting and sealing of the vas deferens.
12. Fertilization is the union of sex cells.
13. The trophoblast is an outer layer of blastula cells necessary for implantation.
14. The placenta is an interlocking structure between the embryo/fetus and mother.
15. Villi are projections, with capillaries, between the embryonic and maternal blood supplies.

Chapter Concepts

1. Development can be described in broad outline but little is known about the details of how cells specialize. What controls which genes are activated or not activated?
2. Important concepts are hormonal signaling, negative feedback, and positive feedback. The increase in estrogen and progesterone shut down the release of FSH (negative feedback). Positive feedback occurs during the uterine contractions at birth.

Practice Test

1. False	11. False	21. d	31. a	41. e
2. True	12. True	22. c	32. d	42. d
3. True	13. False	23. a	33. a	43. g
4. True	14. True	24. d	34. b	44. b
5. False	15. True	25. d	35. b	45. i
6. True	16. True	26. d	36. a	46. h
7. False	17. False	27. c	37. c	47. j
8. True	18. True	28. c	38. b	48. f
9. False	19. False	29. d	39. d	49. c
10. True	20. False	30. c	40. b	50. a

Chapter 29

ANIMAL BEHAVIOR

CHAPTER OVERVIEW

This chapter shows that behavior patterns can be adaptive. They have evolved by natural selection. They range from simple1 to complex responses. Many body systems, including the nervous system, contribute to these responses. Behavior is innate and can be modified by environmental influences.

The details of all aspects of behavior are difficult to document and measure. For example, how does genetics contribute to behavior? Therefore, the principles of this branch of biology will continue to be studied through ongoing research. They will undoubtedly be revised as a result of new studies.

LEARNING OBJECTIVES

After studying this chapter, you should be able to

1. Evaluate the roles of learning and environment on behavior.
2. Cite studies that have enhanced the understanding of behavior.
3. Describe how ethology is a branch of biology.
4. Outline the several social organizations of animal groups.
5. Explain how behavior can be adaptive.
6. State some difficulties in studying behavior.
7. Describe sociobiology.
8. Explain the relationship between genetics and behavior.

CHAPTER OUTLINE

Two Components of Behavior: Instinct and Learning

1. Behavior is a reaction to a stimulus.
2. One idea states that behavior is innate or instinctive.

3. Another idea states that behavior can be modified through learning.

Reflexes and Taxes Are Simple Forms of Behavior

1. A stimulus leads to the reflex response. It is without conscious effort.
2. The knee jerk is an example of a reflex.
3. An animal moves toward or away from a stimulus during a taxis.

Fixed Action Patterns Are Complex, Unlearned Behaviors

1. The FAP is a sequence of complex, unlearned behaviors found within a species.
2. One example is the courtship dance of the male stickleback fish.
3. The are unlearned and apparently inherited.

Learning Alters Behavior through Experience

1. Ethology is the study of animal behavior in natural settings.
2. Associative learning involves conditioned responses.
3. Innate and external factors influence behavior.
4. Several organ systems contribute to responses.
5. Genetics predisposes behavior patterns. They are modified by experience.

Learning Has Genetic Components

1. Studies with fruit flies prove this.

Social Behavior Involves Interaction with Others

1. Animals have a wide range of this type of behavior.
2. There are levels that are adaptive with evolutionary fitness.

Imprinting and the Stages of Social Behavior

1. Lorenz demonstrated this with studies on geese.
2. It has been discovered with other animals.

There Are Three Classes of Social Organization

1. There are aggregates, groups, and societies.
2. Aggretates are random with little internal organization.
3. Groups have organization and division of labor.
4. Societies are the more organized and advanced. Studies of the honeybee hive document this.

Communication Is a Means of Modifying Behavior

1. Pheromones are one means of communication. Some are signaling and some are priming.
2. These chemicals may be present in humans.
3. The bee dance communicates information on food sources. Speed and direction of the dance are meaningful.

Some Animals Have Elaborate Courtship Behaviors

1. Males of some species fight for territory.
2. Studies with gulls show extensive courtship behavior.

Behavior Has an Adaptive Value

1. Behaviors that evolve successfully are adaptive.

Selection and Behavior Are Related

1. Darwin offered examples of this.
2. Studies with widowbirds support this idea.

Social Behavior Contributes to Evolutionary Fitness

1. It does it in several ways, including predator avoidance.

The Evolution of Behavior Is Difficult to Study

1. The communication through the honeybee dance offers some clues.
2. Studies with the removal of eggshells of gulls is another example.

Sociobiology: Genetics and Behavior

1. The application of sociobiology to humans is debatable.
2. Sociobiology has improved the understanding of nonhuman behavior.

Social Structure Offers Advantages to the Individual

1. Aggressive behavior often establishes a dominance hierarchy. This can be adaptive to individual members.

How Did Altruism Originate?

1. There are several explanations.
2. Indirect selection behavior for the benefit of close relatives is favored.
3. Altruism is observed in animal societies such as bees.

Genetics and Human Behavior

1. Not all behavior can be analyzed by simple patterns.
2. Phenotypes undergo continuous change.
3. Genes are inherited and establish the potential for phenotype development.

Single Genes and Human Behavior

1. Some have a simpler analysis such as Huntington disease.
2. Many human behaviors are complex with many gene pairs working.

Genetics and Social Behavior

1. This relationship is unresolved and controversial.
2. Studies are underway to clarify this relationship.

STUDY HINTS

1. Preview the chapter and list all the terms in boldface. After studying the chapter, write definitions for these terms.
2. What approach have you found works best when you study? Try writing practice questions, your own outline, etc.

CHAPTER REVIEW

Key Terms

Define these key terms from the chapter:

1. reflex
2. stimulus
3. taxis
4. FAP
5. ethology
6. conditioned response
7. imprinting
8. aggregate
9. dominance hierarchy
10. altruism

Chapter Concepts

1. What is the relationship between genetics and behavior?
2. What is the relationship between organ system structure and function to behavior?

Practice Test

TRUE/FALSE

___ 1. Behavior is the reaction to a stimulus.

___ 2. A concious effort is required to carry out a reflex.

___ 3. In the knee-jerk reflex a motor neuron signals a skeletal muscle to contract.

___ 4. The movement of flies toward a light source is an example of a geotaxis.

___ 5. A FAP is simply a series of learned behaviors linked together.

___ 6. Fixed action patterns are probably inherited, although learning can modify them.

___ 7. The system of skeletal muscles generates responses in animals.

___ 8. A conditioned response is not a type of associative learning.

___ 9. The term **dunce** refers to a mutated gene in *Drosophila*.

___ 10. The observation of imprinting has been limited to birds.

___ 11. Aggregates usually have a high degree of internal organization.

___ 12. Pheromones are found in a wide range of different kinds of animals.

___ 13. There is only one class of pheromones.

___ 14. The definite identity of pheromones in humans has been established.

___ 15. Honeybees communicate through visual and auditory signals.

___ 16. Animals that live in groups usually do not maintain certain distances between them.

___ 17. A behavior pattern can be viewed as adaptive.

___ 18. Social behavior has probably evolved.

___ 19. Human behavior usually cannot be analyzed from simple Mendelian inheritance patterns.

___ 20. Metabolic by-products do not affect human behavior.

MULTIPLE CHOICE

21. The "nature versus nuture" controversy compares
    a. behavior and physiology.
    b. DNA and RNA.
    c. ecology and evolution.
    d. instinct and learning.
22. A stimulus is a (n)
    a. effector.
    b. environmental change.
    c. receptor.
    d. response.
23. If an animal turns away from light this is an example of
    a. negative geotaxis.
    b. negative phototaxis.
    c. positive geotaxis.
    d. positive phototaxis.
24. The genus name for the fruit fly is
    a. *Apis.*
    b. *Coleoptera.*
    c. *Drosophila.*
    d. *Eohippus.*
25. The discovery of the FAP proved that behavior was
    a. more complex than just innate or learned.
    b. not dependent on the nervous system.
    c. only innate.
    d. only learned.
26. The stickleback is a kind of
    a. bird.
    b. fish.
    c. insect.
    d. reptile.
27. ____________ studied associative learning.
    a. Darwin
    b. Mendel
    c. Pavlov
    d. Wallace
28. The response of an organism is probably least dependent on the ____________ system.
    a. endocrine
    b. integumentary
    c. nervous
    d. skeletal muscle

29. __________ is the chance that an organism will leave more offspring to the next generation compared to other organisms.
    a. altruism
    b. fitness
    c. gene pool
    d. nonprobability
30. Which of the following is not a broad assemblage of social organizations?
    a. aggregate
    b. group
    c. society
    d. taxis
31. Among honeybees, the worker bees are
    a. fertile females.
    b. fertile males.
    c. infertile females.
    d. infertile males.
32. A tactile signal involves
    a. gas molecules.
    b. light.
    c. sound waves.
    d. touch.
33. A pheromone is a __________ signal.
    a. chemical
    b. electrical
    c. light
    d. vibrating
34. In the studies by von Frisch, if the forager bee moves straight down the comb, the recruits fly
    a. at a 45 degree angle to the sun.
    b. at a 90 degree angle to the sun.
    c. directly away from the sun.
    d. directly toward the sun.
35. In the studies by von Frish, the term *waggle* refers to a type of
    a. dance.
    b. hive.
    c. honeybee.
    d. pheromone.
36. Courtship behavior precedes
    a. any kind of taxis.
    b. flight from territory.
    c. loss of social aggregates.
    d. mating of individuals.

37. Tinbergen concluded that eggshell removal by __________ was adaptive.
    a. gulls
    b. robins
    c. sparrows
    d. starlings
38. Altruistic behavior is __________ behavior.
    a. cooperative
    b. geotropic
    c. noncooperative
    d. phototropic
39. __________ are not passed on from parents to offspring
    a. chromosomes
    b. genes
    c. genotypes
    d. phenotypes
40. Huntington disease is an __________ disease in humans.
    a. autosomal dominant
    b. autosomal recessive
    c. X-linked dominant
    d. X-linked recessive

COMPLETION

41. During the knee-jerk reflex, the signaling is carried out by the __________ system in the organism.
42. Movement toward light is a __________ phototaxis.
43. Ethology is the study of animal __________ in natural settings.
44. __________ is the scientist who conducted the best-known historical studies on imprinting.
45. As members of a group have different roles, they perform a division of __________ .
46. __________ pheromones work over a longer time period than signaling pheromones.
47. Intrusion into an animal's individual space usually results in an __________ on the intruder.

48. The binomial name of the common honeybee is ___________ *mellifera.*

49. Agressive behavior is a social group often establishes a ___________ hierarchy.

50. The ___________ is the sum of observable characteristics of an organism.

ANSWERS

Key Terms

1. A reflex is a response to a stimulus without concious effort.
2. A stimulus is an environmental change.
3. A taxis is the turning movment of an animal toward or away from a stimulus.
4. The fixed action pattern (FAP) is a sequence of complex, unlearned behavior found in a species.
5. Ethology is the study of animal behavior is natural settings.
6. A conditioned response is associative learning in which an animal learns to link a stimulus with an event such as feeding.
7. Imprinting is the development of social bonds during a critical time period.
8. An aggregate is social group with members randomly arranged and with little or no social organization.
9. A dominance hierarchy is a series of social roles established in a group.
10. Altruism is behavior that is cooperative and self-sacrificing.

Chapter Concepts

1. Genes that are inherited set the potential for the development of phenotypes. Environmental influences modify these potentials to varying degrees.
2. The organism responds behaviorly through its organ systems. The brain has a center for emotions. A runner uses leg muscles in an event. Blood (circulation) must flow to the brain and muscles. The blood must be oxygenated (respiration).

Practice Test

1. True	8. False	15. True	22. b
2. False	9. True	16. False	23. b
3. True	10. False	17. True	24. c
4. False	11. False	18. True	25. a
5. False	12. True	19. True	26. b
6. True	13. False	20. False	27. c
7. True	14. False	21. d	28. b

29. b
30. d
31. c
32. d
33. a
34. c
35. a
36. d
37. a
38. a
39. d
40. a
41. nervous
42. positive
43. behavior
44. Konrad Lorenz
45. labor
46. priming
47. attack
48. *Apis*
49. dominance
50. phenotype

Chapter 30

POPULATIONS

CHAPTER OVERVIEW

This chapter explains how the population is a major level of organization. It has a life cycle, a growth phase, and an age structure. Numerous factors, biological and physical, regulate this growth. Extinction is a possibility for a population.

Human population growth worldwide is in a log phase. It will reach six billion in this decade. Humans have had a significant impact on other populations on the planet. Some effects have been positive but many have been negative.

Range and density are properties of all populations. Understanding these properties may help to preserve populations of many different species on the Earth.

LEARNING OBJECTIVES

After studying this chapter, you should be able to

1. List the levels of organization, including the population, in the correct hierarchical order.
2. Outline the phases of a population life cycle.
3. Explain some of the factors that control population size.
4. Distinguish some of the different kinds of factors controlling populations.
5. Describe human population growth patterns.
6. Compare the phases of growth and stability of population size.
7. Explain some examples of symbiosis.
8. Explain how populations can become extinct.
9. Discuss some of the impacts humans have had on other populations.
10. Describe the range and density of populations.

CHAPTER OUTLINE

Ecology and the Hierarchy of Life

1. The hierarchy includes population, community, ecosystem, biosphere.
2. Ecology is the study of how organisms interact with each other and their physical environment.

Populations Have a Life Cycle with Three Phases

1. The three phases are growth, stability, and decline.
2. The population often grows rapidly and then levels off. The decline may be a crash.
3. Population dynamics is the study of population interactions.

Population Growth Is Controlled by Several Factors

1. Populations have the potential to increase at an exponential rate.
2. The intrinsic rate of increase equals the birth rate minus the death rate.

Population Growth Potential Is Related to Life History

1. Life history is the sexual maturation and other characteristics of the population.
2. There are two simple extremes ecologists use to explain the trend of population growth.

Age Structure Can Predict Growth Potential

1. Age structure is the relative proportion of individuals in each age group.
2. Pyramid-shaped structures have the potential for rapid growth.

Human Populations Are in a Growth Phase

1. The human population worldwide will reach 6 billion before the year 2000.
2. Advances in technology have decreased the time it takes to add successive billions to the worldwide human population.

Populations Make a Transition from Growth to Stability

1. The population phases of growth to stability fit a logistic growth model.
2. Populations level off in size when they reach their carrying capacity.

Several Basic Controls Govern Population Size

1. Biological interactions include competition, predation, and symbiosis. There are also physical controls.

Resources in the Physical Environment Can Limit Growth

1. Population growth is limited by the law of the minimum.

Competition Is a Biological Factor Controlling Growth

1. The niche is the biological role of the species.

2. Niche overlap can occur.
3. One species can eliminate another by competitive exclusion. Competitive release results for the survivor.

Predators Help Stabilize Populations

1. Carnivores prey on animals.
2. Herbivores prey on plants.
3. Prey species are seldom eliminated because of protective traits, refuges, and prey switching.

Symbiotic Interactions Can Affect Population Size and Stability

1. Both species benefit in mutualism.
2. In parasitism a parasite benefits and the host is harmed.
3. In commensalism one species benefits and the other is not affected.
4. In amensalism one species is inhibited and the other is not affected.

The Real World: Complex Interaction of Population Controls

1. Controls do not work separately.

Physical versus Biological Control Systems in Population Growth Control

1. Physical controls are sometimes density-independent.
2. Biological controls are often major and density-dependent.
3. Both usually affect the growth of a population.

Populations Can Decline and Become Extinct

1. Extinction is the elimination of all individuals of a large group. It can be local or large-scale.
2. Extinction can result from physical or biological factors.
3. There are local examples with human populations.

Human Impact on Population Growth, Stability, and Decline

1. Populations grow as previous limitations are removed.
2. Populations decline as new limitations are added.

Altering Resources Can Selectively Influence Population Growth

1. Eutrophication results from adding P and N, former limiting factors to plant growth.

Pesticides Can Alter Competition

1. Removal of a competitor can allow the other species to grow rapidly.

Removing Predators Can Have Unforeseen Consequences

1. Removal of a predator can lead to rapid growth of the surviving prey species.

Introduction of New Species Can Alter Population Structure

1. A new, exotic species may not have a competitor. Without this control, it can grow rapidly in numbers.

Altering Population Growth Can Be Destructive or Beneficial

1. The growth of killer bee populations in South America is one example of a negative effect.
2. The MSY in agriculture has been a somewhat positive effect.

Intervention Can Cause Population Decline

1. Habitat disruption is a major cause.
2. Species introduction can affect native populations.

Overkills and Secondary Extinctions Can Alter the Biological Environment

1. Species are simply eliminated directly to extinction.
2. A secondary extinction results when one population extinction causes another.

Populations Have a Minimum Viable Size

1. A population may be reduced to a small size that cannot recover.
2. Populations are susceptible to random factors.
3. The MVP indicates the viability of a population.

Range and Density Are Properties of Populations

1. Populations have a geographic range.
2. Their zone of physiological stress is outside their optimum range.
3. Their zone of intolerance is further away from the optimum.

Populations Have Several Dispersion Patterns

1. They are uniform, random, and clumped.
2. Clumped is most common in nature.

Population Ranges Are Dynamic

1. They can be expanded or contracted by natural forces.
2. Human intervention can change the ranges.

STUDY HINTS

1. Preview the chapter by studying the key terms and questions/problems at the end of the chapter before studying.
2. Are you taking a lab course with lecture? What lab experiences are you receiving that you can apply to the study of this chapter? Have you studied populations and ecosystems?

CHAPTER REVIEW

Key Terms

Define these key terms from the chapter:

1. population
2. community
3. ecosystem
4. ecology
5. exponential rate
6. life history
7. age structure
8. carrying capacity
9. competition
10. predation
11. niche
12. mutualism
13. commensalism

14. extinction

15. eutrophication

Chapter Concepts

1. Is it correct to assume that an organism is not isolated from its environment? How is it related?
2. Are humans subject to the same principles of ecology as other species or are they unique?

Practice Test

TRUE/FALSE

___ 1. A population is a broader level of organization than a community.

___ 2. The exponential rate of population growth is very rapid.

___ 3. A population structure with a broad base has the potential for rapid growth.

___ 4. The growth of the human population worldwide was very rapid from 1 A.D. to 1500 A.D.

___ 5. A population size often levels off when it reaches its carrying capacity.

___ 6. Predation is a physical interaction that limits population growth.

___ 7. The niche specifically refers to where an organism lives.

___ 8. Prey switching can prevent the extinction of a prey.

___ 9. Mutualism is a form of symbiosis in which only one species benefits.

___ 10. A community contains more than one species.

___ 11. Extinction is the elimination of all individuals in a group.

___ 12. Eutrophication is runaway, uncontrolled plant growth.

___ 13. Recently the size of the deer population in the United States has been declining.

___ 14. MSY attempts to permit the maximum harvest of food.

___ 15. Species introduction changes the biological environment.

___ 16. Secondary extinctions result from changing the biological environment.

___ 17. A population size must be reduced to zero to cause its extinction.

___ 18. The MVP is a highly accurate measurement of a population size.

___ 19. A population is absent in its zone of intolerance.

___ 20. Members of a population are often uniformly distributed through their geographic range.

MULTIPLE CHOICE

21. Select the most specific level of organization from these choices.
    a. biosphere
    b. community
    c. ecosystem
    d. population
22. Which of the following choices is not a common stage of the population life cycle?
    a. biogenesis
    b. decline
    c. growth
    d. stability
23. The ___________ rate of population increase is the birth rate minus the death rate.
    a. extrinsic
    b. intrinsic
    c. nonsymbiotic
    d. symbiotic
24. A population that has many members that are too young to reproduce has the shape of a(n)
    a. inverted pyramid.
    b. rectangle.
    c. square.
    d. upright pyramid.
25. Human population growth started to become very rapid around the year
    a. 1650.
    b. 1750.
    c. 1850.
    d. 1950.
26. The current human population of the world is about ___________ billion.
    a. two
    b. four
    c. six
    d. eight

27. Which of the following choices is not a biological interaction that can limit population growth?
    a. competition
    b. homeostasis
    c. predation
    d. symbiosis
28. Competitive exclusion refers to the ____________ level of categorizing organisms.
    a. family
    b. kingdom
    c. order
    d. species
29. Commensalism is symbolized as ____________ .
    a. + +
    b. + –
    c. + 0
    d. – 0
30. A human benefits from a microorganism that obtains nutrients in its digestive tract. This is an example of
    a. amensalism.
    b. commensalism.
    c. mutualism.
    d. parasitism.
31. *Tribolium* is the genus name for the
    a. flour beetle.
    b. fruit fly.
    c. honeybee.
    d. wasp.
32. Nitrogen and ____________ are often limiting factors for plant populations.
    a. calcium
    b. copper
    c. phosphorous
    d. selenium
33. Increased plant growth in a stream can deplete the ____________ in the water and kill the fish.
    a. carbon dioxide
    b. carbon monoxide
    c. oxygen
    d. ozone
34. An example of competitive release is
    a. eutrophication.
    b. game releases.

c. nutrient pollution in lakes.
d. poisoning of insect pests.

35. "Killer bees" are the result of
    a. budding.
    b. cloning.
    c. hybridization.
    d. sterilization.
36. Fast-growing mussels from Asia have been a particular problem for other populations in the
    a. Everglades.
    b. Great Lakes.
    c. Mississippi River.
    d. Pacific Ocean.
37. MSY is based on the premise of an __________ -shaped population graph.
    a. A
    b. E
    c. S
    d. X
38. Small-scale disturbances of habitat disruption affect the __________ species.
    a. fastest-moving
    b. least susceptible
    c. most susceptible
    d. slowest-moving
39. The extinction of the dodo bird had a direct impact on a population of
    a. fish.
    b. mammals.
    c. reptiles.
    d. trees.
40. The most common pattern of population distribution is
    a. clumped.
    b. extinction.
    c. tolerance.
    d. uniform.

COMPLETION

41. __________ is the study of how organisms interact with each other and with their physical environment.
42. Population __________ is the study of factors and interactions that affect the growth, stability, and decline of a population.

43. By the __________ growth model the exponential growth of a population is limited by some factor.
44. By competitive __________ one species eliminates another species through competition.
45. –0 is the symbolism for __________ , one type of symbiosis.
46. Predation is one example of a density- __________ factor that can control population size.
47. A drought is an example of a density- __________ factor that can control population size.
48. The Japanese beetle first reached the United States by reaching the state of __________ .
49. By MSY the maximum rate of a population experiencing sigmoid growth occurs at a fraction of __________ of its carrying capacity.
50. __________ disruption is currently the major cause of extinction of populations in the United States.

ANSWERS

Key Terms

1. A population is a group of organisms of the same species.
2. A community is all the different species grouped in an area.
3. An ecosystem includes the community and nonliving factors in an area.
4. Ecology is the study of how organisms interact with each other and their physical environment.
5. The exponential rate is the rapid rate of population growth.
6. The life history of a population refers to the sexual maturation, age of death, and other characteristics.
7. The age structure refers to the relative proportion of individuals in each age group of a population.
8. The carrying capacity is the point where a population levels off in size.
9. By competition two species require and compete for the same resource.
10. In predation a predator feeds on a prey species.
11. The niche is the biological role of a species.
12. Mutualism is a form of symbiosis in which both species benefit.

13. Commensalism is a form of symbiosis in which one species benefits and the other is not affected.
14. By extinction all individuals are eliminated from a group.
15. Eutrophication is runaway plant growth.

Chapter Concepts

1. The key word in ecology is *interactions.* An organism depends on other organisms for nutrition. It may be a source of nutrition for another species. It numbers are regulated by predation, commensalism, and physical factors.
2. Humans are subject to the same principles of population dynamics as all species. They have phases of population change. Their numbers can be regulated. They affect other populations.

Practice Test

1. False	14. True	27. b	40. a
2. True	15. True	28. d	41. ecology
3. True	16. True	29. c	42. dynamics
4. False	17. False	30. c	43. logistic
5. True	18. False	31. a	44. exclusion
6. False	19. True	32. c	45. amensalism
7. False	20. False	33. c	46. dependent
8. True	21. d	34. d	47. independent
9. False	22. a	35. c	48. New Jersey
10. True	23. b	36. b	49. one-half
11. True	24. d	37. c	50. habitat
12. True	25. a	38. c	
13. False	26. c	39. d	

☙ Chapter 31 ❧

COMMUNITIES AND ECOSYSTEMS

CHAPTER OVERVIEW

This chapter informs the student about the associations of organisms with other organisms and with the physical component of their environment. Communities have populations with certain dispersal and abundance patterns. The biomes are the worldwide terrestrial communities. Two kinds of aquatic biomes have distinct characteristics. Diversity and stability are properties of all kinds of communities. These properties can experience disturbances.

Communities are not static but everchanging through succession. Species replacement occurs in this type of evolution.

In ecosystems energy flows and matter cycles. Human intervention can affect the efficiency of these processes.

LEARNING OBJECTIVES

After studying this chapter, you should be able to

1. Describe community structure.
2. Describe the distribution and species richness patterns in communities.
3. Compare the terrestrial biomes for similarities and differences.
4. Explain how climate and altitude affect terrestrial biomes.
5. Describe the two kinds of aquatic biomes.
6. Explain diversity and stability in communities.
7. Outline the pattern of community succession.
8. Explain how disturbances affect community diversity.
9. Outline examples of the flow of energy and cycling of materials in ecosystems.

10. Discuss how human activity can affect energy flow and the cycling of materials in ecosystems.

CHAPTER OUTLINE

What Are Communities and Ecosystems?

1. A community is all of the populations that live in a given area.
2. An ecosystem is the community plus the physical environment.

Community Structure Depends on Population Dispersion Patterns

1. Two patterns are open structure and the rarity of species.

Most Communities Have an Open Structure

1. A closed community is a discrete unit.
2. Populations in open communities are distributed randomly.
3. Most communities are open.
4. The exception is where there is an abrupt transition in the physical environment.

Open Communities Are Not Tightly Knit Groups of Organisms

1. Communities are not integrated as in a superorganism.

Most Species Are Rare in Communities

1. Populations often have peak abundance at the center of their range.
2. Usually there are only a few dominant populations in a community.
3. The general cause is resource partitioning.

Communities Are Grouped into Classifications

1. Communities are terrestrial or aquatic.
2. A biome is a large-scale category of similar communities.

There Are Six Terrestrial Biomes

1. The tropical rain forests have the highest species diversity. They are found near the equator.
2. The temperate forest includes much of the eastern United States.
3. Grassland are in temperate and tropical regions with reduced rainfall.
4. The desert has dry conditions and wide temperature ranges.
5. The taiga is a northern coniferous forest.
6. The tundra is between the taiga and regions of northern permafrost.

Climate and Altitude Affect Terrestrial Biomes

1. There is control by an altitudinal gradient and latitudinal gradient.
2. Cooler conditions cause aridity except in deserts.
3. Tropical forests are in warm equatorial areas.

There Are Two Types of Aquatic Biomes

1. Aquatic environments have conditions generally less harsh than on land.
2. The marine biome has benthic and pelagic zones.
3. Pelagic organisms are planktonic and nektonic.
4. There is a crucial euphotic zone.
5. The freshwater biome lake and pond have a stratification of zones.
6. There is a mixing of zones in the fall and spring.

Communities Have Two Properties: Diversity and Stability

Community Diversity Is Measured by Species Richness

1. Diversity refers to the different kinds of organisms in the community.
2. Species richness is influenced by a latitudinal diversity gradient.
3. Three interrelated factors help produce this gradient: environmental stability, community age, and length of growing season. Nutrient availability is another influence.

Community Stability Is Related to Structure

1. Stability is the ability of a community to persist unchanged.
2. Species-rich ecosystems tend to be stable.

Communities Change Through Time

1. Geological time focuses over long time spans than ecological time.

Community Succession Is a Form of Species Replacement

1. It is the sequential replacement of species in a community.
2. New species enter the community. Others become extinct.
3. A pioneer community reaches a climax community.
4. There are many examples in forest communities with trends including decreasing productivity.
5. Biomass increases in latter stages.

Community Evolution Occurs over Long Time Spans

1. It can involve millions of years.
2. Current communities generally have more species.
3. Species packing occurs on the ocean floor.

Disturbance Can Cause Communities to Lose Their Diversity

1. Human activity usually simplifies communities.
2. Monoculture is one example.
3. Communities become stressed with fewer species.

Ecosystems and Community Function

1. Energy flows and matter cycles.

Energy Flows through Ecosystems

1. Energy flows through the food chain, food web, and food pyramid.
2. The food chain is a series of feeding organisms.
3. The food web is a more accurate representation.
4. The food pyramid relates the energetics of producers and consumers.
5. The number of steps in the food chain are usually limited.

Ecosystem Productivity Begins at the Base of the Food Pyramid

1. NPP is the rate at which producer biomass is made.
2. Aquatic ecosystems are the most productive.
3. It is most productive along the shorelines.
4. NSP is the rate which consumer and decomposer biomass is produced.

Human Activity Can Disturb Energy Flow and Productivity

1. Humans redirect a lot of energy away from other species.

Matter Cycles through Ecosystems

1. C is one element circulating through a biogeochemical cycle.
2. Undisturbed natural ecosystems cycle elements very efficiently.
3. Cycling is generally faster and more efficient in tropical ecosystems.

Human Activity Can Disturb Cycling of Matter

1. Excess output can occur.
2. Slash and burn agriculture is one example.
3. Disturbances can occur by excess input.

STUDY HINTS

1. This chapter has extensive content and many key terms. Make mastery of vocabulary a major goal of your study.
2. Can you visit a local terrestrial ecosystem? Try to at least make some rough estimates about its diversity and distribution of organisms. Can you outline one of its food chains? These firsthand observations can add insight to the study of this chapter.

CHAPTER REVIEW

Key Terms

Define these terms.

1. community

2. ecotone

3. biome

4. aridity

5. benthic

6. pelagic

7. stratification

8. diversity

9. species richness

10. stability

11. community succession

12. biomass

13. monoculture

14. food chain

15. producer

Chapter Concepts

1. Diversity is adaptive for a community. Support this position.
2. Does community succession also occur in aquatic ecosystems? Offer an example.

Practice Test

TRUE/FALSE

___ 1. A community is all the different populations in a certain area along with the physical environment.

___ 2. Ecotones are sharp boundaries between closed communities.

___ 3. Most of the species in a community are very abundant.

___ 4. The taiga is an aquatic biome.

___ 5. Tropical rain forests are near the equator in each hemisphere of the Earth.

___ 6. A temperate forest can develop into a grassland through a lack of rainfall.

___ 7. Angiosperms are the dominant plant in the taiga.

___ 8. With increasing altitude the climate generally tends to become cooler.

___ 9. Aridity refers to the moisture content of the air.

___ 10. The euphotic zone is the part of the marine biome where light penetrates.

___ 11. Oxygen levels in the water tend to increase in the deeper zones of a lake.

___ 12. Species richness tends to increase in latitudes away from the equator.

___ 13. Ecological time focuses on community events over thousands of years.

___ 14. Succession has been most fully documented in aquatic communities.

___ 15. In a community, biomass is the total amount of living tissues.

___ 16. Human activity often tends to decrease the species diversity in a community.

___ 17. The primary consumers in a community are the photosynthetic organisms.

___ 18. The open ocean is an ecosystem that is not highly productive.

___ 19. Matter cycles in an ecosystem whereas energy flows in an ecosystem.

___ 20. The cycling of matter is often slow and inefficient in tropical ecosystems.

MULTIPLE CHOICE

21. Select the broadest level of organization from these choices.
    a. community
    b. ecosystem
    c. organism
    d. population
22. Populations in open communities are distributed more or less
    a. by clusters.
    b. in aggregates.
    c. randomly.
    d. uniformly.
23. Ecologists often apply the term "superorganism" to
    a. communities.
    b. ecosystems.
    c. individual organisms.
    d. the biosphere.
24. Which of the following statements is not generally true about communities?
    a. A combination of species is present.
    b. Most species are rare.
    c. There is resource partitioning.
    d. They are inflexible to the environment.
25. A best example of a secondary factor affecting communities is the
    a. amount of rainfall.
    b. availability of nutrients.
    c. temperature.
    d. soil conditions.
26. The rainfall range for a tropical rain forest annually is usually __________ inches.
    a. 10 to 20
    b. 40 to 60
    c. 80 to 160
    d. 300 to 400
27. Which of the following statements is not correct about the desert biome?
    a. It has dry conditions.
    b. Plants store water in succulent tissues.
    c. It receives less than 10 inches of rainfall annually.
    d. Its temperatures are very stable.

28. Permafrost is found most often in the
    a. grassland.
    b. temperate forest.
    c. taiga.
    d. tundra.
29. If moisture moves across a continent from the west, the rain shadow will develop most on the ___________ side of the inland mountain ranges.
    a. east
    b. north
    c. south
    d. west
30. A benthic organism dwells
    a. on the bottom of oceans.
    b. on the flat land of deserts.
    c. in the open, surface water.
    d. in trees.
31. The most obvious difference between ponds and lakes is
    a. location.
    b. pH level.
    c. salinity.
    d. size.
32. The mixing of the upper and lower layers of water in a lake is the ___________ overturn.
    a. spring and fall
    b. spring and summer
    c. winter and fall
    d. winter and summer
33. Generally, the species diversity of an aquatic community increases with the increasing ___________ of the water.
    a. depth
    b. pH
    c. sunlight
    d. temperature
34. A "cascade collapse" most directly threatens the ___________ of a community.
    a. diversity
    b. energy flow
    c. position
    d. size
35. The initial colonizing species is the ___________ community.
    a. climax
    b. grassland

c. pioneer
d. successive

36. In the process of preparation in community succession, each stage of a community
   a. changes from the climax to other pioneer stages.
   b. changes from the climax to pioneer stages.
   c. sets conditions for its own demise.
   d. sets conditions for its own success.
37. Planting only corn in a field for a year is an example of
   a. community diversity.
   b. monoculture.
   c. species packing.
   d. succession.
38. In the food chain of alga—minnow—sunfish—bass—bear, the most useful energy is available to which consumer?
   a. bass
   b. bear
   c. minnow
   d. sunfish
39. Of the following choices, the food __________ does not represent an energy flow through an ecosystem?
   a. chain
   b. niche
   c. pyramid
   d. web
40. Humans use about __________ percent of the land food pyramid base on the Earth.
   a. 10
   b. 20
   c. 40
   d. 80

COMPLETION

41. A __________ is a large-scale category that includes many communities of similar characteristics.
42. Community __________ is the sequential replacement of species over time.
43. The ecological __________ refers to how an organism functions in an ecosystem.
44. __________ agriculture consists of cutting down and burning trees.

MATCHING

Match, by letter, each description to the correct term. Each choice is used once.

___	45. desert	a. Evergreens are the dominant land plant.
___	46. grassland	b. The major animals are grazers.
___	47. taiga	c. It is the most complex biome.
___	48. temperate forest	d. The topsoil is frozen all year.
___	49. tropical forest	e. Oak and maple trees are dominant.
___	50. tundra	f. Cacti store water here.

ANSWERS

Key Terms

1. A community is the set of all populations that inhabit a certain area.
2. An ecotone is a sharp boundary between communities.
3. A biome is a large-scale category that includes many similar communities.
4. Aridity is the degree of dryness in a community.
5. Benthic refers to bottom-dwelling organisms in a marine biome.
6. Pelagic refers to organisms in the open water of the marine biome.
7. Stratification of water refers to the layers that develop in lakes and ponds seasonally.
8. Diversity refers to the different kinds of organisms in a community.
9. Species richness is the number of different species in a community.
10. Stability is the ability of a community to persist unchanged.
11. Community succession is the sequential replacement of species in a community.
12. Biomass is the total weight of living tissue in a community.
13. A monoculture is a community with only one species.
14. A food chain is a series of organisms feeding in a certain order.
15. A producer generates food for all organisms in the food chain.

Chapter Concepts

1. Diversity offers checks and balances in a community. If, for example, a species has several predators it can still be controlled if one predator is eliminated. This is adaptive for preserving the entire community.
2. A deep pond could gradually become a shallow pond. The shallow pond could become a marsh and eventually pass through the stages of a terrestrial ecosystem.

Practice Test

1. False
2. True
3. False
4. False
5. True
6. True
7. False
8. True
9. True
10. True
11. False
12. False
13. False
14. False
15. True
16. True
17. False
18. True
19. True
20. False
21. b
22. c
23. a
24. d
25. b
26. c
27. d
28. d
29. a
30. a
31. d
32. a
33. a
34. a
35. c
36. c
37. b
38. c
39. b
40. c
41. biome
42. succession
43. niche
44. slash and burn
45. f
46. b
47. a
48. e
49. c
50. d

∞ Chapter 32 ∞

BIOSPHERE AND HUMAN IMPACT

CHAPTER OVERVIEW

This chapter presents information that should enlighten the student about the future of the Earth and its life forms. Human impact has been substantial in the lithosphere, hydrosphere, and atmosphere. As the biosphere depends on these three spheres, it has also been affected.

Problems in the biosphere include the effects of global warming and acid deposition, loss of species diversity, and threatened food production. Some problems are economic as well as biological. Solutions are being sought and implemented. Their success will determine the future of the biosphere into the next century.

LEARNING OBJECTIVES

After studying this chapter, you should be able to

1. Distinguish among the lithosphere, hydrosphere, and atmosphere.
2. Explain the relationship of the other three spheres to the biosphere.
3. List some effects that are disrupting the biosphere.
4. Describe the greenhouse effect.
5. Describe the threat to the ozone layer.
6. Describe the origin and effects of acid rain.
7. Compare the sources of water pollution.
8. Outline the changes currently occurring in the tropical regions.
9. Discuss species extinction and why it is important to stop it.
10. Discuss the loss of genetic diversity on the planet.

11. List some solutions being implemented to save the biosphere.
12. Discuss the accompanying economic problems that result from destruction of the biosphere.

CHAPTER OUTLINE

The Biosphere and Its Physical Environment

1. The biosphere is the sum of all living matter on the Earth.
2. Most life is close to the Earth's surface.

The Lithosphere Is the Outer Layer of the Earth

1. Land covers about 30% of the surface of the Earth.
2. It currently undergoes a tectonic cycle with the upwelling of magma.
3. An upwelling forms a divergent plate boundary.
4. A convergent plate boundary is formed under new material.
5. Land masses go through continental drift.
6. The tectonic cycle has produced many environmental changes.

The Hydrosphere Is the Water at or Near the Earth's Surface

1. It is the most abundant substance in living organisms.
2. The hydrologic cycle depends on evaporation, transpiration, and precipitation.
3. The hydrosphere is only a small part of the Earth's total water.

The Atmosphere Envelops the Earth

1. It is 78% nitrogen and 21% oxygen.
2. It is densest near the surface of the Earth.

A Living Planet: Integration of the Spheres

1. A element such as C passes through all three spheres.
2. There are quantitative differences for most elements between the biosphere and the nonliving part of the environment.

Biogeochemical Cycles Connect the Biosphere with the Physical Spheres

1. All four spheres may form a superorganism.

The Biosphere Is Being Disrupted at Many Levels

1. Human activity can be a major source of disruption.

Global Changes in the Atmosphere

1. Trace gases make up an important 1%.
2. There are two sinks for atmospheric carbon.
3. More carbon dioxide has recently been added to the atmosphere.
4. The main source is the burning of fossil fuels.

Carbon Dioxide Emissions May Contribute to Global Warming
1. This occurs through the greenhouse effect.
2. There are other greenhouse gases.
3. Global warming is occurring regionally.

The Ozone Layer Is Being Destroyed
1. Ozone filters high-energy radiation from the sun.
2. It is being destroyed by the release of CFCs.

Acid Deposition and Smog Are Local Atmospheric Changes
1. Acid deposition comes from acid rain.
2. The source of smog is fuel combustion.
3. Most energy in the U.S. is from burning fossil fuels.

Water Pollution Is Altering the Hydrosphere
1. Acid rain from sulfur pollution pollutes waterways.
2. Global warming could melt polar ice caps.
3. Water pollution arises from municipal sewage, agriculture, mining, logging, and industrial toxins.

Are We on the Verge of a Sixth Mass Extinction?
1. More than 50% of the species living in an area must become extinct for this to occur.
2. The last one was 65 million years ago.

Development is Accelerating in Tropical Regions
1. One-half of the species could be extinct by the middle of the next century.
2. 80% of the species of the world are in the tropical rain forests.
3. Coral reefs are also diverse and vulnerable to pollution.

Some Species Are More Susceptible to Extinction
1. Species with limited habitats are susceptible.
2. Tropical insects and plants are also susceptible.

Why Try to Save Species from Extinction?
1. Some of the reasons are economic.

Crop Plants Have Lost Their Genetic Diversity
1. Four plants provide most of the food for the world.
2. Low diversity makes organisms more vulnerable.
3. Genetic diversity has biological and economic advantages.

Managing Ecosystems Can Have Economic Benefits
1. Many medicines originated from tropical plants.
2. Many plants cannot be domesticated.

Ecosystems Are Environmental Support Systems
1. Plants make oxygen.
2. Microorganisms purify the environment.
3. Cycles are balanced.

Taking Action to Reduce Species Extinction and Biosphere Destruction
1. Some initial steps have been taken.

Research and Description of Species
1. Biodiversity must be measured.
2. There are other types of diversity: genetic, species, and ecosystem.
3. All of these measurements are difficult.

Establishment of Preserves
1. There has been some success with preserves.
2. A problem is acquiring enough land for a preserve.

Laws Protecting Endangered Species
1. One is the Endangered Species Act of 1973.
2. It has generated some controversy.

Breeding in Captivity
1. It is less desirable but sometimes is the only choice.

Reduce Socioeconomic Causes of Extinction
1. Rapid human population growth is a major cause of species extinction.
2. Industrialization, pollution, and resource depletion must be slowed to reduce population decline of other species.

STUDY HINTS

1. Select the study approach that has worked best for you throughout this book. Use it in this chapter and try to perfect it as you approach the study of other courses.
2. Continue to use the key terms and the questions/problems at the end of the chapter for preview and then for a thorough summary.

CHAPTER REVIEW

Key Terms

Define these key terms from the chapter:

1. biosphere

2. lithosphere

3. hydrosphere

4. atmosphere

5. Gaia

6. sink

7. greenhouse effect

8. smog

9. mass extinction

10. gene bank

11. ecosystem diversity

12. debt-for-nature swap

Chapter Concepts

1. Explain how the spheres of the Earth integrate.
2. The issues of biology and economics are often inseparable. How does this develop?

Practice Test

TRUE/FALSE

___ 1. The lithosphere is the solid outer layer of the Earth.

___ 2. Magma is molten rock.

___ 3. Oxygen is the most common gas in the atmosphere.

___ 4. Oxygen is the most abundant element in the Earth's crust.

___ 5. The oceans are a sink to absorb carbon dioxide from the atmosphere.

___ 6. The addition of carbon dioxide in the atmosphere provides an avenue for the escape of heat from the Earth.

___ 7. Methane is a greenhouse gas.

___ 8. There is a trend that the amount of ozone is increasing in the atmosphere of the Earth.

___ 9. Sulfur dioxide is released into the atmosphere from the burning of fossil fuels.

___ 10. The major source of energy in the United States is from fossil fuels.

___ 11. Global warming could cause the sea levels to rise.

___ 12. More than 20% of the species that have existed on the Earth are still alive.

___ 13. Life began on the Earth about 2.0 billion years ago.

___ 14. Tropical regions contain an estimated 80% of the species on the Earth.

___ 15. Cockroaches have undergone significant evolution over the last 100 million years.

___ 16. Small plant species in the tropics are currently experiencing high rates of extinction.

___ 17. Only four different species of plants provide most of the food for the world.

___ 18. Plants in the tropical rain forest serve as a source of new drugs.

___ 19. Species diversity is the same as ecological diversity.

___ 20. Populations that are bred in captivity usually thrive and reproduce at high rates.

MULTIPLE CHOICE

21. The hydrosphere covers about ____________ percent of the Earth.
    a. 30
    b. 50
    c. 70
    d. 90
22. Select the incorrect statement about the lithosphere of the Earth.
    a. It has about 10 large plates.
    b. It has numerous small plates.
    c. The crust is one layer of it.
    d. The plates are on a nonmolten base.
23. The human body is about ____________ percent water.
    a. 45
    b. 65
    c. 86
    d. 95
24. Select the second most abundant element in the Earth's crust.
    a. aluminum
    b. oxygen
    c. silicon
    d. sodium
25. Select the most abundant element in living organs.
    a. carbon
    b. hydrogen
    c. nitrogen
    d. oxygen
26. Limestone is
    a. calcium carbonate.
    b. calcium phosphate.
    c. sodium bicarbonate.
    d. sodium carbonate.
27. Most of the energy of the world is derived from
    a. fossil fuels.
    b. geothermal power.
    c. nuclear reactors.
    d. photovoltaic cells.
28. Computer models show that global warming of the Earth will cause
    a. all areas to become drier.
    b. all areas to become wetter.
    c. some areas to become drier and some will become wetter.
    d. the areas to hold moisture uniformly worldwide.

29. Ozone
    a. filters high-energy radiation from the Sun.
    b. is produced from nitrogen atoms.
    c. is produced from phosphorous atoms.
    d. produces high-energy radiation from the Sun.
30. Each of the following is a common source of CFCs except
    a. air conditioners.
    b. refrigerators.
    c. solvents.
    d. UV radiation sources.
31. Which of the following is a major source of water pollution?
    a. agriculture
    b. industries
    c. municipal sewage
    d. recreation
32. In a mass extinction, at least ___________ percent of all species living become extinct.
    a. 20
    b. 40
    c. 50
    d. 75
33. The rain forests cover about ___________ percent of the Earth's surface.
    a. 7.5
    b. 12.5
    c. 18
    d. 21
34. Many recent mammal extinctions have occurred
    a. in lakes and streams.
    b. in the oceans.
    c. on islands.
    d. on mainlands.
35. In the Irish potato famine, the potatoes were affected by a
    a. bacterium.
    b. fungus.
    c. protozoan.
    d. virus.
36. Select the cereal crop.
    a. barley
    b. cassava
    c. chickpea
    d. groundnut

37. Presently about ____________ plant species supply about 95 % of the nutrition of the world.
    a. 30
    b. 60
    c. 90
    d. 120
38. Select the legume.
    a. bean
    b. maize
    c. potato
    d. sorghum
39. About ____________ percent of the land surface of the Earth has been set aside for preserves.
    a. 1
    b. 6
    c. 10
    d. 15
40. The major industry in the rain forests is
    a. forestry.
    b. entertainment.
    c. pharmaceuticals.
    d. tourism.

COMPLETION

41. The lithosphere of the Earth is not static but is changing or ____________ .

42. ____________ is the evaporation of water from plants.

43. The number of spheres of Gaia is ____________ .

44. ____________ is a gas and reactant of photosynthesis.

45. The accumulation of carbon dioxide in the atmosphere has caused the ____________ effect.

46. A decrease in the ozone layer in atmosphere has been associated with an increased rate of human ____________ cancer.

47. Smog is caused by fuel ____________ .

48. The ____________ regions of the Earth have high species diversity.

49. Gene ____________ are used to save the seeds, spores, and other genetic materials of organisms.

50. Rubber trees are a forest-sustainable ___________ .

ANSWERS

Key Terms

1. The biosphere is the sum of all living matter on the Earth.
2. The lithosphere is the solid outer layer of the Earth.
3. The hydrosphere is all of the water at or near the surface of the Earth.
4. The atmosphere is the envelop of gases surrounding the Earth.
5. Gaia is the name for a proposed superorganism consisting of the four spheres of the Earth.
6. A sink can absorb carbon dioxide from the air.
7. In the greenhouse effect, a layer of carbon dioxide builds up in the atmosphere and traps escaping heat, thus warming the Earth.
8. Smog is a mixture of air pollutants, including sulfur and nitrogen oxides.
9. Mass extinction occurs if more than 50% of the species living become extinct.
10. A gene bank saves genes in seeds, spores, and sperm.
11. Ecosystem diversity is the number of local ecosystems or communities.
12. In a debt-for-nature swap a country sets aside pristine land for a preserve in exchange for reduction in the financial debt owed to another country.

Chapter Concepts

1. Follow the cycle of carbon through the three nonliving spheres of the environment as well as the biosphere. All elements of life have a component in the nonliving part. Their availability to life depends on their availability and cycling.
2. Biological resources have economic value. Study the examples of food production from angiosperms and animals. Ecosystems are valuable for recreation as well. They also have other value.

Practice Test

1. True	11. True	21. c	31. d	41. dynamic
2. True	12. False	22. d	32. c	42. transpiration
3. False	13. False	23. b	33. a	43. four
4. True	14. True	24. c	34. c	44. carbon dioxide
5. True	15. False	25. d	35. b	45. greenhouse
6. False	16. True	26. a	36. a	46. skin
7. True	17. True	27. a	37. a	47. combustion
8. False	18. True	28. c	38. a	48. tropical
9. True	19. False	29. a	39. a	49. banks
10. True	20. False	30. d	40. b	50. resource